Hoppe

FERNS

AND

FERN ALLIES

OF

CALIFORNIA

BY

STEVE J. GRILLOS

ILLUSTRATED BY RITA WHITMORE

UNIVERSITY OF CALIFORNIA PRESS

BERKELEY, LOS ANGELES, LONDON 1971

UNIVERSITY OF CALIFORNIA PRESS

BERKELEY AND LOS ANGELES, CALIFORNIA

UNIVERSITY OF CALIFORNIA PRESS, LIMITED
LONDON, ENGLAND

© 1966 BY THE REGENTS OF THE UNIVERSITY OF CALIFORNIA

SECOND PRINTING, 1971
ISBN: 0-520-00519-8
LIBRARY OF CONGRESS CATALOG CARD NUMBER: 65-19247

PRINTED IN THE UNITED STATES OF AMERICA

CONTENTS

County map of California.

INTRODUCTION

What are ferns? How do they differ from the common seed-bearing plants which are familiar to all of us? What are the unusual "spots" that form on the bottomside of many fern leaves? These questions and many others are raised by the layman who sees these interesting plants in their natural environment. For those who work closely with nature's landscape, ferns and their relatives offer a unique opportunity for investigation.

Ferns are for the most part land-inhabiting. They possess well-formed roots, stems, and leaves. They are structured like the gymnosperms (conifers, ginkgoes, etc.) and flowering plants in that they form a well-developed vascular system that serves for the conduction of water, mineral salts, and foods. All plant types forming such a system are placed under the plant phylum *Tracheophyta*. Although the ferns are considered to be primarily land plants, some require free water, especially during parts of their reproductive cycle.

In studying ferns, one soon recognizes that these plants differ from the seed-bearers in several ways; however, the most significant difference is in the method of reproduction. The gymnosperms and flowering plants produce a multicellular reproductive structure, commonly known as the seed, from which new plants form; whereas the ferns reproduce mainly by spores, which are one-celled, asexual reproductive units. The significance of the spore in relation to the

[5]

total life cycle of ferns is described later. Grouped with the ferns are other kinds of unusual plants such as the club-mosses and horsetails, which have similarities with the ferns, but differ somewhat in their structural features. For convenience, in various sections of this guide, the term "fern" is also used to refer to these kinds of plants.

Fossil records indicate that ferns have been a part of the earth's vegetation for millions of years. The living forms are a mere handful compared to the countless number that once flourished. The present literature indicates that approximately 10,000 species have been collected, identified, and recorded. This is a relatively small number compared to the more than 325,000 species of seed-bearing plants. These plants have always been of keen interest to students of nature's landscape because they exhibit many structural differences, have unique life-cycle patterns, and are successful in nearly all types of environmental conditions.

The most successful ferns are limited to tropical habitats, where many have become treelike, extending 40 feet or more in the air. In the temperate regions they are smaller and are widely distributed. They grow on dry, rocky outcrops at lower and higher elevations; in open mountainous places; in shady woodland meadows; in ponds, lakes, and swampy places either floating or partially emersed; along the roadsides; and even within cities, more commonly in vacant lots.

Some eighty-six species of ferns are native to California. There is some disagreement on the number, depending upon whether certain plants have been given species or variety status. Fern nomenclature (naming of plants) is a subject on which even the specialists disagree. It is not within the scope of this guide to discuss the position at which certain plants are to be placed. This guide has been prepared to familiarize you with the ferns most commonly encountered in field work, and therefore does not include all the eighty-six or so species. A more complete survey of the fern population is given in the latest flora for the state, *A California Flora* (Berkeley and Los Angeles: University of California Press, 1959), by P. A. Munz and D.D. Keck. To identify any of the nonnative ferns you may refer to manuals which have been so designed, as for example, *Manual of Cultivated Plants* (New York: The Macmillan Co., 1949), by L.H. Bailey.

The ferns discussed in this book are divided into eight groups, commonly called families (see the Family Key on p. 14). These in turn are subdivided into several subcategories and finally to the scientific name which includes the genus and species. For a discussion of the method and purpose of naming plants, see *Introduction to the Natural History of the San Francisco Bay Region*, p. 18 (Berkeley and Los Angeles: University of California Press, 1959), by Arthur C. Smith or *Introduction to the Natural History of Southern California*, p. 29 (Berkeley and Los Angeles: University of California Press, 1965), by Edmund C. Jaeger and Arthur C. Smith.

As an example, the common Maiden-Hair Fern (*Adiantum capillus-veneris*) is classified as follows:

Kingdom: *Plant*
  Subkingdom: *Embryophyta*
    Phylum: *Tracheophyta*
      Subphylum: *Pteropsida*
        Class: *Filicinae*
          Order: *Filicales*
            Family: *Polypodiaceae*
              Genus: *Adiantum*
                Species: *capillus-veneris*

To make it easier for the reader working from this guide to Munz's *Flora*, since it is the latest comprehensive flora and because there is confusion in fern names, you will find under certain species descriptions a cross reference. This field guide includes the following items of interest:

1. How to identify a fern.
2. Life cycle of a typical fern.
3. Key to families of ferns and fern allies.
4. Family and generic descriptions; generic keys.
5. Descriptions of species with illustrations; species keys; habitat, elevation, range, and importance to man, if any.
6. A section on activities which can add greatly to your knowledge and pleasure.
7. Glossary of terms.
8. Check list of ferns and fern allies of California.

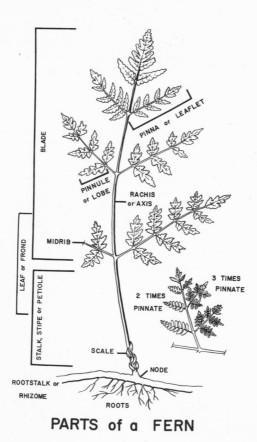

PARTS of a FERN

[8]

HOW TO IDENTIFY A FERN

In order to identify a fern, follow the guide lines listed below.

1. Collect a complete specimen. The plant should have a root-stalk, several leaves, and sori.
2. Examine your specimen and refer to the illustration on page 8 to determine the structural features.
3. Compare your specimen with the description in the key to the families and the marginal illustrations. A key is a device, set up in outline form, whereby a combination of structural features for larger groups of plants and specific plants is given. Check though the outline and eliminate features which do not fit the given plant, and, by means of the features which are present, you can identify the plant. Botanical manuals supply keys for the various plant groups, e. g., classes, subclasses, families, genera, and species.
4. Select the family, go to the page indicated, and read the family description to see if you have correctly placed it in this group.
5. Check the key to the genera, if there is one, and select the the proper genus; then read the description and compare it with your specimen.
6. Finally, identify the species by using the key provided under the genus description. Check your plant with the species description and the illustration. It is a good idea to make a careful record of the habitat, distribution, etc., of the plant when you are in the field, since this information is helpful in identification.
7. If your determination is incorrect, go back through and try again. You may wish to double-check your identification by referring to more extensive botanical publications. (See suggested references.)

[9]

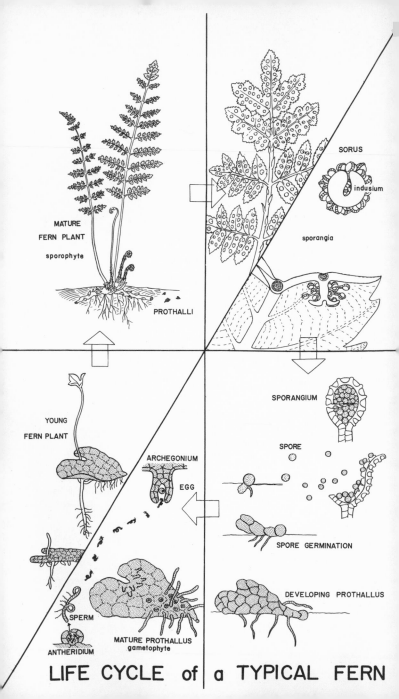

MATURE
FERN PLANT
sporophyte

PROTHALLI

SORUS

indusium

sporangia

SPORANGIUM

SPORE

SPORE GERMINATION

DEVELOPING PROTHALLUS

YOUNG
FERN PLANT

ARCHEGONIUM

EGG

SPERM

ANTHERIDIUM

MATURE PROTHALLUS
gametophyte

LIFE CYCLE of a TYPICAL FERN

LIFE CYCLE OF A TYPICAL FERN

All ferns possess an alternation of generations—a condition in which the complete life cycle involves two generations. The sporophyte generation (spore-bearing plant) is the most conspicuous in having true roots, stems, and leaves, and is independent, except for a short period in its development when it is nutritionally dependent upon the gametophyte. The gametophyte or prothallus (gamete-bearing plant) is small, relatively simple, and does not develop roots, stems, and leaves. Most ferns produce spores of one kind; a few produce spores of two kinds, commonly known as microspores and megaspores. A spore germinates and develops into the gametophyte, or sexual generation, which produces the sex cells. The fertilized egg of the gametophyte develops into the sporophyte without the formation of a seed.

STRUCTURAL FEATURES

A typical fern, such as a Sword Fern, Shield Fern, or a Wood Fern, has a characteristic horizontal underground stem, commonly called a rootstalk or rhizome. The leaves develop from buds which differentiate at the nodes on the upper surface or the stem; the roots form from opposite points along the lower surface. The stems function in supporting the roots and leaves; the roots anchor the plant in the soil and absorb water and minerals. The majority of our temperate species have the rhizome type of stem. Some ferns, like those growing in the tropics, produce a stem which is erect and trunklike.

The leaves (fronds) are usually flat, green, expanded, and supported by a slender stalk, sometimes referred to as the stipe or petiole. Fern leaves vary in form, size, and function. The blade, the expanded part of the leaf, may be dissected into smaller segments

[11]

known as pinnae (singular, pinna). If these in turn are divided, the smaller segments are called pinnules. A few ferns have leaves of two kinds. One kind functions in reproduction, forming spores; these leaves are referred to as fertile leaves, or sporophylls. The other kind, sterile leaves, do not form spores and are entirely photosynthetic (food-making) in function. In some ferns there is little difference in form between fertile and sterile leaves. Both types make food; the fertile leaf, in addition, forms spores. In other species the fertile leaf has become modified so greatly that it shows little resemblance to the sterile type. The majority of ferns produce one type of leaf which functions in food-making and spore-production. The spore-forming bodies are confined to specific areas.

REPRODUCTION

The spore-forming bodies develop on the underside of the blade, or in special leaves which produce these bodies but generally do not make foods. The spore bodies, known as sporangia (singular, sporangium), are usually borne in groups called sori (singular, sorus), or, more popularly, "fruit dots," which appear as brownish spots or streaks. The sori are usually covered with a thin membranous tissue developing from the lower epidermis of the blade. This is called the indusium (plural, indusia). In a few ferns the margins of the blade may curl over and form this covering. The distribution of sporangia on the leaf surface varies considerably in different genera and species: the sporangia may (1) cover much of the lower surface, (2) be grouped in sori growing in a definite relationship with the veins, or (3) grow only along the margins or edges of the blade or segments.

Each sporangium consists of an expanded outer part (the capsule) and a short stalk. The capsule is a slightly compressed, almost circular, biconvex saclike structure, with a row of thick-walled cells (the annulus) about two-thirds of the way around the outer edge. At the end of the annulus is a group of thin-walled lip cells. In the early stages of development, the capsule contains specialized cells, known as spore mother cells, each of which divides to form four cells. Each cell contains a reduced number of chromosomes and matures into a spore. One sporangium produces approximately 64 spores. As the sporangium matures, it ruptures at its weakest point, where the lip cells are situated, and releases the spores. The annulus, owing to its unique structure, is sensitive to humidity changes and can move in a springlike fashion that helps to scatter the spores.

The spores, after being released from the capsule, fall to the ground and germinate, if conditions are favorable, into heart-shaped structures known as prothalli (singular, prothallus). These structures are the gametophyte plants. The prothalli are green, simple in structure, one to several cells in thickness around the outer edges and somewhat thicker toward the center. They rarely exceed more than one-quarter of an inch in diameter. The prothalli of many ferns grow horizontally on the soil surface; however, in a number of species they are frequently found below the surface. From the underside develop rootlike structures, termed rhizoids, which function in much the same way as do the roots of the sporophyte. The male and female sex organs differentiate on the lower surface and are simple in structure.

The male sex organ, or antheridium, consists of a single layer of cells toward the outside; the center section produces approximately 32 sperm cells. The archegonium, the female sex organ, has a short neck and an inflated base in which a single egg develops. When the sex organs are ripe, the antheridium ruptures, releasing the sperm, which swim to the neck of the archegonia and travel down to the bases and unite with the eggs. Only one sperm unites with each egg. The first cell formed after fertilization, the zygote, has one set of chromosomes from the sperm cell and one set from the egg. The zygote undergoes repeated cell division and develops into an embryo sporophyte comprising a foot, primary root, stem, and leaf. Early in its development, the embryo obtains nutrition from the prothallus until it develops a leaf which will form the necessary foods. This young sporophyte then grows into the mature fern plant and the cycle is complete.

The sporophyte generation of a fern plant includes all stages of the life cycle beginning with the zygote, formed from the fusion of the two sex cells, to the point at which the spore mother cells undergo division to produce spores in the sporangia. The sporophyte is the dominant and more complex plant in the total life cycle. The gametophyte generation begins at the spore stage and ends at the time of fertilization. This phase is relatively short in terms of the total life cycle.

KEY TO THE FAMILIES OF FERNS AND FERN ALLIES

Stems jointed, hollow; leaves reduced to
    toothed sheaths around joints .... Horsetail Family
                      (*Equisetaceae*), p. 16

Stems not jointed, solid; leaves well
    developed, and not as above.
        Plants primarily aquatic, unattached, free-
            floating .... Water Fern Family
                          (*Salviniaceae*), p. 23

        Plants semi-aquatic or terrestrial, attached
          by roots, not free-floating.
             Leaves 4-foliolate (clover-like) or
                filiform; sporangia develop within hard,
                ovoid to globose sporocarps .. Pepperwort Family
                            (*Marsileaceae*), p. 24

[14]

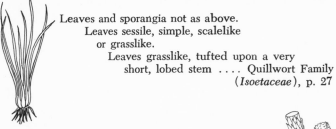

Leaves and sporangia not as above.
    Leaves sessile, simple, scalelike
    or grasslike.
        Leaves grasslike, tufted upon a very
            short, lobed stem .... Quillwort Family
                (*Isoetaceae*), p. 27

Leaves awl-like or scalelike,
    ½ of an inch long or less, imbricated
    in several ranks.
        Leaves ⅛ to ½ of an inch long; spores
           of one kind .... Club-Moss Family
              (*Lycopodiaceae*), p. 31

        Leaves less than ⅛ of an inch long;
           spores of two kinds .. Spike-Moss Family
              (*Selaginellaceae*), p. 32

Leaves stalked (petiolate), usually
    broad, often variously lobed or
    divided, rarely linear or scalelike.
        Sporangia borne in spikes or
        panicles at terminal parts
        of sporophylls, sessile,
        usually opening by a transverse
        slit .... Adder's Tongue Family
            (*Ophioglossaceae*), p. 38

        Sporangia borne on undersurface
        of leaves, short-stalked, often
        covered by membranous covering
        (indusium) .... Fern Family
          (*Polypodiaceae*), p. 44

HORSETAIL FAMILY (*Equisetaceae*)

Horsetails are rushlike herbaceous plants generally found in wet soils. They have an extensive underground system of blackish, jointed rootstalks, bearing roots at the joints. Aerial stems are present; they are usually erect, sometimes branching, dimorphic (of two kinds: sterile and fertile) or uniform, grooved, cylindrical, jointed, hollow between the joints, traversed by hollow tubes, and often overlaid with silica deposits. The leaves are in whorls at the joints and reduced to variously toothed sheaths. Sporangia are ovoid and develop on the underside of sporophylls, which form short cones (strobili) terminating the fertile stems. Spores are all of one kind, green, and bear hygroscopic, elongated bands (elaters).

The prothalli are small, green, and variously lobed: they grow in damp places.

This family includes only one genus.

SCOURING - RUSHES (*Equisetum*)

The structural features are as given for the family.

The generic name is from the Latin *equus*, horse, and *seta*, bristle, signifying the horsetail-like appearance of many of the branched species.

KEY TO THE SPECIES

Cones tipped with a rigid point; stems usually perennial (evergreen).

    Sheaths nearly or quite as broad as long, cylindrical, usually ashy with 2 black bands .... *E. hyemale*

    Sheaths much longer than broad, distended above, somewhat funnel-shaped, green, the lower ones with a narrow black band .. *E. laevigatum*

Cones more or less rounded at tip; aerial
stems annual, not surviving colder
periods of year.
Aerial stems all alike, green.
Mature stems rough, with cross-bands of
silica; tufts of branched sterile
stems develop at base of
fertile ones .... *E. funstoni*
Mature stems smooth, with or without
cross-bands of silica; tufts of branched
sterile stems lacking .... *E. kansanum*
Aerial stems dimorphic, the sterile ones
much-branched, green, the fertile ones unbranched, short-lived, pale or brownish.
Sterile stems 4 to 24 inches long, 6- to 14-
grooved; sheaths with 8 to 12 teeth .... *E. arvense*
Sterile stems 1½ to 4 (or 8) feet long, 20- to 40-
grooved; sheaths with 20 to 30 teeth .. *E. telmateia*

Common Scouring-Rush (*Equisetum hyemale*)

Structural Features: Aerial stems usually perennial, all similar,
usually unbranched, stout, green 2 to 4 feet long, 16- to 48-
angled, the ridges roughened by 1 or 2 rows of silica tubercles.
Leaves nearly or quite as broad as long, cylindrical, usually
ashy with 2 black bands; their teeth brown with whitish margins, quite persistent. Cones less than ½ of an inch to 1¼
inches long, nearly sessile, tipped with a rigid point.
Habitat: Moist alluvial soils and streams in hills and mountains,
from 100 to 5,000 feet.
Range: Coast Ranges, inward from Humboldt County to southern California.

This species is variable in some of its characteristics
and therefore has been separated by many writers
into several varieties, based usually on the number of
rows of tubercles on the ridges and the number of
keels on the leaves: *E. hyemale* var. *californicum; E.
hyemale* var. *robustum; E. hyemale* var. *intermedium.*

If you are camping and failed to bring along scour-
ing materials, stems from this plant may be used to
scour your pots and pans. The rough siliceous surface
makes it desirable for this use. The stems have been
used also to polish floors and other woodwork.

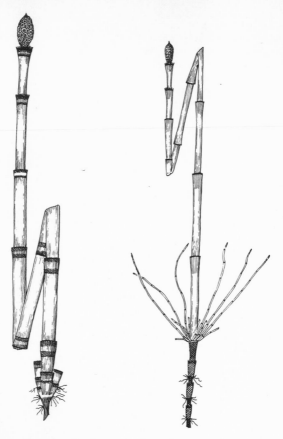

Common Scouring-Rush   Braun's Scouring-Rush

Braun's Scouring-Rush (*Equisetum laevigatum*)
Structural Features: Aerial stems usually perennial, all similar, commonly unbranched, erect, tufted, pale green, 12 to 18 inches long, 20- to 30-angled, mostly smooth. Leaves much longer than broad, slightly dilated upward, green; the lower ones with a narrow black band, their teeth often deciduous. Cones less than ½ to ¾ of an inch long, tipped with a rigid point.
Habitat: Damp alluvial thickets and sandy banks, below 6,500 feet.
Range: Sierra Nevada and northward.

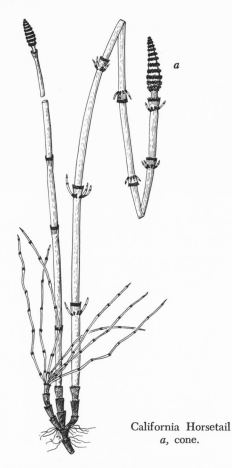

California Horsetail
a, cone.

California Horsetail (*Equisetum funstoni*)

Structural Features: Aerial stems annual, all alike, slender, tufted, 1½ to 6 feet long, 20- to 30-angled, very rough, the ridges with many sharply projecting cross-bands of silica. Leaves elongated, funnel-form, green; their teeth deciduous, the persistent base forming a very narrow blackish band. Cones slender, ½ of an inch to 1¼₆ inches long, more or less rounded at the tip.

Habitat: Common in moist places, in partial shade, below 8,000 feet.

[19]

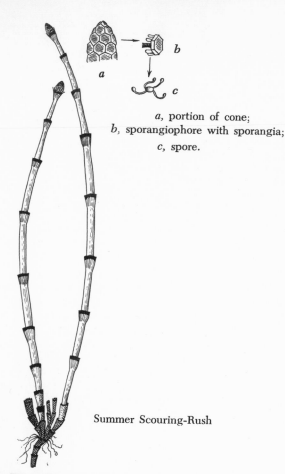

a, portion of cone;
b, sporangiophore with sporangia;
c, spore.

Summer Scouring-Rush

Range: San Diego County northward to Humboldt, Butte, and Siskiyou counties.

This species is also variable, and a number of growth forms have been named: *E. funstoni* var. *caespitosum; E. funstoni* var. *nudum; E. funstoni* var. *ramosum.*

Summer Scouring-Rush (*Equisetum kansanum*)
Structural features: Aerial stems annual, all similar, single or clustered, erect, usually without branches or basal tufts, pale green, 1 to 3 feet long, 15- to 30-angled, smooth or slightly roughened.

[20]

Leaves dilated upward, green with one black band at base; their teeth black with whitish margins, partly deciduous. Cones oblong-ovoid, $\frac{3}{16}$ of an inch to $1\frac{1}{4}$ inches long, blunt at tip.
Habitat: Wet meadows, bogs, and alluvial thickets, below 6,500 feet.
Range: Owens Valley, Inyo County; Santa Rosa Mountains, Riverside County; cismontane California.

Common Horsetail (*Equisetum arvense*)
Structural features: Aerial stems of 2 kinds (dimorphic), prostrate to erect, annual. Sterile stems, green, 4 to 24 inches long, much-branched, 6- to 14-grooved, roughish; their sheaths with about 12 hyaline-margined teeth; their branches numerous, in dense whorls, solid, sharply 3- to 4-angled. Fertile stems pale or brownish, 2 to 10 inches long, unbranched, short-lived; their sheaths pale, with 8 to 12 large brownish lanceolate teeth. Cones lance-ovoid, $\frac{3}{4}$ of an inch to $1\frac{1}{8}$ inches long, blunt at tip.
Habitat: Sandy wet soils or swamps, below 10,000 feet.
Range: Modoc County southward to Inyo County; cismontane California.

This species is extremely variable in habit and size. The Common Horsetail is considered to be an annoying weed in several parts of the state. It has been consumed by cattle and horses, particularly in the younger stages of plant growth, but has been found to be poisonous to horses.

Giant Horsetail (*Equisetum telmateia* var. *braunii*)
Structural features: Aerial stems of 2 kinds (dimorphic), erect, short-lived. Sterile stems white or pale green, $1\frac{1}{2}$ to 4 (or 8) feet long, branched, 20- to 40-grooved, smooth, their sheaths with 20 to 30 broadly hyaline-margined teeth; their branches numerous, in dense whorls, solid, 4- to 6-angled. Fertile stems whitish or brownish, 12 to 18 inches long, almost always unbranched, short-lived; their sheaths brown, loose, membranous, with 20 to 30 teeth. Cones stout-pedunculate, 2 to 3 inches long, blunt at tip.
Habitat: Swampy gullies and along streams, below 5,000 feet.
Range: Cismontane California.

The stems of this plant have been processed and utilized to make a diuretic tea (one which increases the flow of urine).

[21]

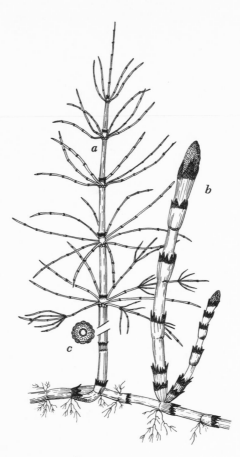

Common Horsetail

a, sterile stem; *b*, fertile stem; *c*, cross-section of stem.

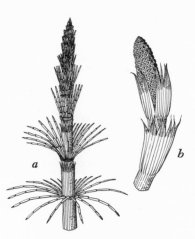

Giant Horsetail
a, sterile stem; *b*, fertile stem.

WATER FERN FAMILY (*Salviniaceae*)

Water Ferns, world-wide in distribution, are small mosslike floating plants found only in aquatic places; they seem to prefer quiet fresh water. The stems are essentially flat, with many roots beneath, and two rows of small leaves on the upper surface. Megaspores and microspores are present and can easily be differentiated on the basis of their size. Both spore types develop on stalks within large receptacles, termed sporocarps.

We have only the genus *Azolla*.

DUCKWEED FERNS (*Azolla*)

The stems are pinnately branched and covered with small, overlapping, 2-lobed leaves. The sporocarps develop in pairs on the underside.

The generic name is from the Greek *azo*, to dry, and *ollupi*, to kill, signifying the rapid death of the plant when it is removed from the water.

We have only one species.

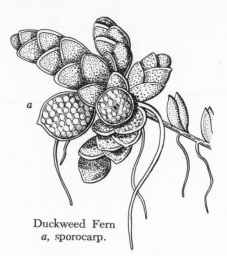

Duckweed Fern
a, sporocarp.

Duckweed Fern (*Azolla filiculoides*)

Structural features: Plants green or reddish, ¼ of an inch to 1 inch broad and as long, easily breaking apart. Leaves ovate, less than ½ of an inch long.

Habitat: Scattered localities throughout state, from sea level to 1,000 feet.

Range: Modoc, Mendocino, San Francisco, Santa Clara, Kern, Stanislaus, Inyo, Ventura, San Bernardino, Riverside and San Diego counties.

During the winter months these plants produce reddish pigments in their leaves and stems, imparting a reddish tinge to the water.

PEPPERWORT FAMILY (*Marsileaceae*)

The Pepperworts usually grow in wet places, around ponds, streams, and meadows. These small, herbaceous, perennial plants have slender, creeping, underground stems, which are frequently branched, and have one or more leaves developing from their nodes. Leaves are either 4-foliolate (clover-like) or simple and petiolate. Sporangia are encased within hard, ovoid to globose, stalked bodies (sporocarps), which

[24]

develop from near the bases of the leafstalks or consolidated with them. Megaspores and microspores develop within the same sporocarp.

The prothalli are usually few-celled and either project from or are included within the spore wall.

Two genera are represented in California.

KEY TO THE GENERA

Leaves 4-foliolate (clover-like); sporocarps ovoid
   or bean-shaped .... *Marsilea*
Leaves simple, filiform; sporocarps globose .... *Pilularia*

PEPPERWORTS OR CLOVER FERNS (*Marsilea*)

The leaves are compound, with the blade divided into equal parts (leaflets) at the end of a long stalk. The sporocarps are ovoid or bean-shaped and have teeth near the base. Upon germination, the sporocarps burst at length into valves, emitting a gelatinous, wormlike mass bearing the large number of sporangia.

The genus was named for Count Luigi Fernando (1658-1730), whose name was Aloysius Marsigli (in Latin, Marsilius). Marsigli was an early Italian patron of botany.

KEY TO THE SPECIES

Upper tooth of sporocarp sharp, conspicuous;
   megasporangia 12 to 20 in each sorus .... *M. vestita*
Upper tooth of sporocarp a rounded projection,
   or sometimes obsolete; megasporangia 6 to 9
   in each sorus ......;..................... *M. oligospora*

Hairy Pepperwort (*Marsilea vestita*)

Structural features: Nodes of stems clothed with hairs. Leaves few to many, arising from nodes, 2 to 8 inches long. Leaflets broadly cuneate, entire, deciduously hairy, ¼ of an inch to 1⅛ inches long, nearly as broad. Sporocarps solitary, at first densely hairy, ⅛ to ½ of an inch long; ⅛ to ¼ of an inch wide; upper tooth rather long, sharp, frequently curved. Each sorus with 12 to 20 megasporangia.

Habitat: Edges of ponds, streams, and swamps from near sea level upward.

Range: Scattered localities in San Diego County northward through Central Valley to Siskiyou and Modoc counties.

[25]

Hairy Pepperwort
a, M. oligospora, sporocarp; *b, M. vestita,* sporocarp;
c, a germinated sporocarp.

In many areas in the United States this species has been reported as a weed. Occasionally it has been used as a feed by horses.

Nelson's Pepperwort (*Marsilea oligospora*)

This species is rather limited in its distribution. It has been reported in Tulare County and probably ranges northward. The species are questionably distinct, with *M. vestita* the more abundant one. *M. oligospora* is more tufted, has permanent soft hairs over the surface of the leaflets, the upper tooth of the sporocarp is rounded or sometimes obsolete, and each sorus produces from 6 to 9 megasporangia.

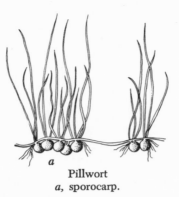

Pillwort

a, sporocarp.

PILLWORTS (*Pilularia*)

Pillworts are very small, inconspicuous plants of muddy places. Their leaves are simple, filiform, without a blade, ¾ of an inch to 1½ inches long, and develop from various points along the stem. The sporocarps are globose, short-stalked, and axillary.

The generic name is from the Latin *pilula,* a little ball, referring to the globose shape of the sporocarp. We have only one species.

Pillwort (*Pilularia americana*)

Structural features: Leaves one or more developing at each node, the nodes frequently rooting beneath. Sporocarps about ⅛ of an inch in diameter, attached laterally to a short, descending stalk, and usually with 3 cavities.

Habitat: Prefers heavy soils in muddy places, below 6,000 feet.

Range: Siskiyou and Modoc counties and southward to San Diego County.

QUILLWORT FAMILY (*Isoetaceae*)

This family contains small, usually erect, grasslike herbs submerged or partially emersed in streams, ponds, or shallow lakes. The stem, commonly called a corm, is short, fleshy, 2-to 3-lobed, and has many roots developing from the base. The leaves are in basal tufts, 1 inch to 11 inches long, narrow, somewhat 3- to

4-angled, spoonlike at the base, and develop in a spiral arrangement around the stem. Internally, the leaves have longitudinal air chambers, a central vascular cylinder, sometimes with peripheral strands of supporting cells, and often with openings (stomates) in the epidermis. The sporangia develop on the upper surface of the expanded base of the leaf; they are solitary, large, orbicular or ovoid, and usually covered by a thin membranous tissue known as the velum. Besides the velum, there is a small triangular extension of epidermal tissue above each sporangium. This is known as the ligule. Two types of sporangia are formed: megasporangia and microsporangia. The former contain the large, globular, whitish megaspores which produce the female sex organs; the latter contain the small, obliquely oblong and triangular, brownish microspores in which are produced the male sex organs.

The prothalli are simple in structure and develop within the spore wall.

There is only one genus of world-wide distribution.

QUILLWORTS (*Isoetes*)

The structural features are as given for the family.

The generic name is from the Greek *isos*, same, and *etos*, year, probably because most of the species retain their leaves for the better part of the year.

KEY TO THE SPECIES

Plants partially emersed or terrestrial; peripheral
  strands usually present.
    Stems 2-lobed; velum not well developed (about one-third
      complete); leaves 2 to 11 inches long ...... *I. howellii*
    Stems 3-lobed; velum well developed (entirely
      complete).
        Leaves 12 to 60, 1¼ to 8 inches long,
          peripheral strands usually 3 .... *I. nuttallii*
        Leaves 6 to 14, 1⅛ to 2¼ inches long,
          peripheral strands lacking .... *I. orcuttii*
Plant mostly submerged and covered by water
  most of the year; peripheral strands not
  present in leaves.

Leaves without stomates; megaspores marked
  with irregular ridges on basal faces ... *I. occidentalis*
Leaves with a few stomates; megaspores
  variously warted.
    Megaspores marked with small spines. Trinity
      and Siskiyou counties ....*I. muricata*
    Megaspores marked with low tubercles or
      wrinkles. San Bernardino Mountains, Sierra
      Nevada, and northward .... *I. bolanderi*

Howell's Quillwort (*Isoetes howellii*)

Structural features: Medium-sized plant with a 2-lobed stem.
Leaves bright green, slightly spreading, mostly 10 to 30, 2 to
11 inches long, with many stomates, usually with 4 peripheral
strands. Sporangia frequently spotted, oval, covered one-third
or less by velum. Ligule narrow, elongate-triangular, equaling
sporangium. Megaspores white, faintly marked with tubercles
and distinct crests.
Habitat: Borders of lakes and ponds, below 9,000 feet.
Range: Mesas near city of San Diego inland to San Bernardino
Mountains; Coast Ranges; Sierra Nevada northward to Modoc
and Del Norte counties.

Nuttall's Quillwort (*Isoetes nuttallii*)

Structural features: Terrestrial plant with a slightly 3-lobed
stem. Leaves light green, erect, 12 to 60, 1¼ to 8 inches long,
with numerous stomates, usually with 3 peripheral strands.
Sporangia oblong or oval, completely covered by velum. Megaspores mostly grayish, usually densely tuberculate.
Habitat: Damp places, along banks of streams and rivers, below
9,000 feet.
Range: Mesas near San Diego, northward through Coast Ranges
and Sierra Nevada.

Orcutt's Quillwort (*Isoetes orcuttii*)

Structural features: Terrestrial plant with a slightly 3-lobed
stem. Leaves erect to slightly spreading, 6 to 14, 1⅛ to 2¼
inches long, with stomates, without peripheral strands. Sporangia orbicular to slightly elongate, completely covered by velum.
Megaspores gray (or brownish when wet), small, smooth, shiny,
often sparsely mealy.
Habitat: Winter pools on mesas, at lower elevations.
Range: San Diego County to Contra Costa and Sacramento
counties.

[29]

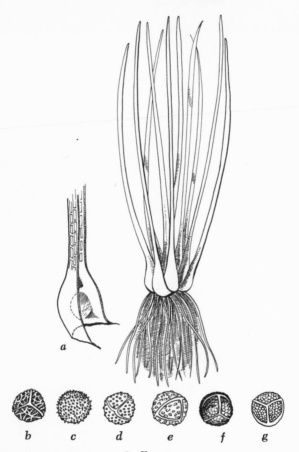

Quillwort

a, enlarged leaf base with sporangium; *b*, *I. occidentalis*, mega-spore; *c*, *I. muricata*, megaspore; *d*, *I. bolanderi*, megaspore; *e*, *I. howellii*, megaspore; *f*, *I. orcuttii*, megaspore; *g*, *I. nuttallii*, megaspore.

Western Quillwort (*Isoetes occidentalis*)

Structural features: Submerged species with a 2-lobed stem. Leaves dark green, rigid, 9 to 30 (to 60), 2 to 8 inches long, without stomates and peripheral strands. Sporangia almost or-bicular, about one-third covered by velum. Megaspores cream-colored, marked with irregular ridges.

Habitat: Mountainous meadows, along shores of ponds and lakes, from 5,000 to 11,000 feet.
Range: Siskiyou County southward through Sierra Nevada.

Braun's Quillwort (*Isoetes muricata*)
Structural features: Submerged species with a 2-lobed stem. Leaves green or reddish-green, straight or recurved, 10 to 18, 3 to 6 inches long, with a few stomates toward tip, without peripheral strands. Sporangia spotted, oblong, over one-half covered by velum. Megaspores white, marked with small spines.
Habitat: Lakes and ponds, at about 8,000 feet.
Range: Siskiyou and Trinity counties.

Bolander's Quillwort (*Isoetes bolanderi*)
Structural features: Small submerged plant with a deeply 2-lobed stem. Leaves bright green, soft, erect, 6 to 25, 2 to 5 inches long, with a few stomates, usually without peripheral strands. Sporangia covered about one-third by velum. Megaspores white, sometimes bluish, marked with low tubercles or wrinkles.
Habitat: Shallow mountain lakes or lakelets and ponds, from 5,000 to 11,500 feet.
Range: San Bernardino Mountains and northward through the Sierra Nevada, Cascade, and Klamath Ranges.

Isoetes bolanderi shows some degree of variability in the length of the leaf, and for this reason the variety *pygmaea* has been established. The leaf in this form is from ¾ of an inch to 1¼ inches long. It has been reported growing at higher elevations near streams and lakes in the southern region of the Sierra Nevada.

CLUB-MOSS FAMILY (*Lycopodiaceae*)

Club-mosses are mosslike, perennial, evergreen terrestrial plants. The stems are usually branched, erect or creeping, and leafy nearly throughout. The leaves are many, simple, ⅛ to approximately ½ of an inch long, 1-nerved, lanceolate, and usually crowded into several ranks. Sporangia are 1-celled, solitary at the base of typical leaves, or of modified leaves (sporophylls), forming specialized cones (strobili) at the tips of erect stems. The spores are yellow, many, small, globose, smooth to variously roughened, and all of one kind.

[31]

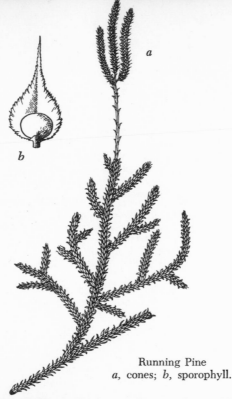

Running Pine
a, cones; *b*, sporophyll.

The prothalli are fleshy, with or without chlorophyll, and usually develop below the soil surface.

We have only one genus.

CLUB-MOSSES (*Lycopodium*)

The structural features are as given for the family.

The generic name is from the Greek *lykos*, a wolf, and *podium*, a foot, so named because the root resembles a claw.

We have two species, neither of which is widely distributed; however, they are common beyond our borders. *L. clavatum* is most frequently encountered in the field; *L. inundatum* has been reported in only one locality, Arcata, Humboldt County.

Running Pine (*Lycopodium clavatum*)

Structural features: Main stems 12 to 24 inches long, prostrate, wide-creeping, branching horizontally, giving rise to many leafy, pinnately divided, ascending branches. Leaves of ultimate branches crowded, incurved-spreading, entire or denticulate, tipped with a bristle which is usually deciduous. Cones solitary or clustered, narrowly cylindric, 1 to 2 inches long, developing on ends of slender stalks (2 to 3 inches long).

Habitat: Forming dense masses on trees in moist coniferous woods and boggy places, at about 500 feet.

Range: Humboldt County and northward.

The spores are sold at drug counters under the name *Lycopodium*, and have been used in acoustical experiments, as powders in fireworks, and for coating surfaces of pills to prevent their sticking together. The entire plant has been used to some extent in the designing of Christmas decorations.

SPIKE-MOSS FAMILY (*Selaginellaceae*)

Spike-mosses are low, mosslike, leafy, evergreen terrestrial plants. The stems are slender, freely branched, and horizontal to erect. The leaves are scalelike, many, simple, less than ⅛ of an inch long, 1-nerved, subulate to lanceolate, all alike or 2 or more sizes, and crowded into several ranks. The sporangia are 1-celled, subglobose, solitary, developing near the base on the ordinary leaves, or on modified leaves (sporophylls) that form more or less 4-angled cones (strobili) at the tips of the branches. The spores are of 2 kinds: megaspores, 1 to 4, usually in the lower part of the cone; microspores, many, minute, reddish or orange, usually in the upper part of the cone.

Male and female prothalli are formed and are confined within the spore walls. The male prothallus develops from the microspore and forms male sex cells: the female prothallus develops from the megaspore and forms female sex cells.

We have only one genus.

[33]

SPIKE-MOSSES (*Selaginella*)

The structural features are as given for the family.

The generic name, a diminutive of *Selago*, is a classical name for some species of *Lycopodium*.

Approximately 11 species have been reported for California.

Most of these forms are distinguishable only on the basis of careful examination. Abrams and Munz and Keck (see suggested references) include comprehensive discussions of all 11 types. Included here are 5 common types; the other species are listed to show only their distribution.

KEY TO THE COMMON SPECIES

Plants ascending or erect, rooting only at base .... *S. bigelovii*
Plants creeping or trailing, rooting throughout.
    Leaves in 4 ranks (2 planes). Northern
        California .... *S. douglasii*
    Leaves arranged radially in many ranks.
        Stems not at all dorsoventral. Marin
            County and northward .... *S. wallacei*
        Stems dorsoventral (almost flat).
            Leaves with a terminal bristle. Mountains
                from Shasta to Tulare County .... *S. hanseni*
            Leaves without a terminal bristle.
                Desert .... *S. eremophila*

Bushy Selaginella (*Selaginella bigelovii*)

Structural features: Stems slender, ascending or erect, 2 to 8 inches long, shortly branched, rooting only at or near base. Leaves appressed-imbricate, narrow-lanceolate, ciliate at margins, with a short bristle at tips. Cones erect at tips of short lateral branches, ⅛ to ½ of an inch long; their sporophylls similar to typical leaves, except that they are deltoid-ovate.
Habitat: Exposed, dry rocky slopes and ridges, below 7,000 ft.
Range: Coast Ranges from Sonoma County and southward; Kern and Tulare counties in Sierra Nevada; along western edges of the deserts; Santa Barbara Islands.

Douglas' Selaginella (*Selaginella douglasii*)

Structural features: Stems prostrate, wide-creeping, 3⅛ to 16 inches long; rooting throughout, leafy throughout; their branches alternately arranged, pinnately divided several times, 2 to 6 inches long. Leaves rigid, distant, spreading, less than ⅛ of an

[34]

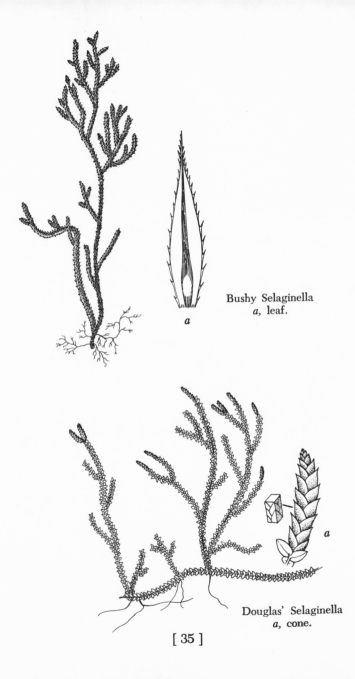

Bushy Selaginella
a, leaf.

Douglas' Selaginella
a, cone.

inch long, arranged in 2 planes, with those of upper plane half as long as those of lower, and shortly pointed. Cones many, sharply quadrangular, slightly curved, ⅛ to ½ of an inch long; their sporophylls closely imbricate, cordate-acuminate, sharply keeled.

Habitat: Moist, shady rocky slopes, below 6,000 feet.
Range: Scattered localities in northern California.

Wallace's Selaginella (*Selaginella wallacei*)

Structural features: Stems creeping, forming loosely cushion-like tufts or mounds, 2 to 8 inches long, freely branched, rooting sparsely throughout; their branches numerous, ascending, not at all dorsoventral, ¼ of an inch to 1½ inches long. Leaves with texture of writing paper, rigidly appressed-imbricate on all sides, mostly oblong-linear, ciliate at margins, with terminal bristles. Cones numerous, slightly curved, slender, ¼ of an inch to 1⅛ inches long.

Habitat: Shady, damp places and dry rocky slopes, below 5,000 feet.
Range: Siskiyou and Humboldt counties southward along coast to Marin County.

Hansen's Selaginella (*Selaginella hanseni*)

Structural features: Stems prostrate, creeping, forming cushion-like tufts or mounds, 2 to 10 inches long, branching, rooting throughout; their branches strongly dorsoventral, curving toward upper part of main stem. Leaves linear-lanceolate, dark ashy or purplish with age, ciliate at margins, with a bristle at tips. Cones numerous, at ends of branchlets, slightly curved, $\frac{3}{16}$ to ⅜ of an inch long; their sporophylls deltoid-ovate, acuminate, ciliate on either margin.

Habitat: Exposed rocky slopes, from 1,000 to 5,500 feet.
Range: Foothills of western side of Sierra Nevada from Tulare County northward to Shasta County; Santa Lucia Mountains.

Desert Selaginella (*Selaginella eremophila*)

Structural features: Stems prostrate, close-creeping 2 to 4¾ inches long, much branched, rooting sparsely throughout; their branches short, strongly dorsoventral. Leaves arranged radially in many ranks, broadly lanceolate, acutish, brownish below, bright green above, conspicuously ciliate at margins, coiling over stems when they become dry. Cones many, slightly curved, ¼ to ⅜ of an inch long; their sporophylls similar to typical leaves.

Habitat: Crevices of exposed, rocky slopes, below 3,000 feet.
Range: Canyons along western border of Colorado Desert.

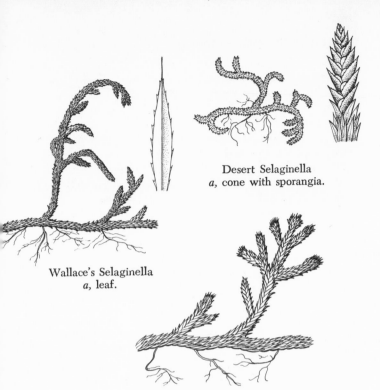

Desert Selaginella
a, cone with sporangia.

Wallace's Selaginella
a, leaf.

Hansen's Selaginella

Pygmy Selaginella (*S. cinerascens*). San Diego County.

Oregon Selaginella (*S. oregana*). Del Norte and Humboldt counties.

Alpine Selaginella (*S. watsoni*). At higher altitudes in Sierra Nevada.

Bluish Selaginella (*S. asprella*). San Bernardino, San Gabriel, San Jacinto, Santa Ana, Laguna Mountains (San Diego County).

Rocky Mt. Selaginella (*S. densa* var. *scopulorum*). Siskiyou County.

Mojave Selaginella (*S. leucobryoides*). Providence Mountains and Panamint Range, Mojave Desert.

ADDER'S TONGUE FAMILY (*Ophioglossaceae*)

These plants are primarily terrestrial, but often grow in moist places. They all have a short, fleshy, underground stem, and many fibrous roots. The leaves are from one to several, but mostly one, erect, simple or variously divided, sessile or stalked, consisting of a simple dissected sterile blade (here simply called the leaf) and a stalked spore-bearing spike or panicle (known as the sporophyll). Both of these structures develop on a common stalk, called the petiole. The basal stalk is sheathing at the base and contains the bud for the next growing season. The sporangia develop in rows at the margins and are large and globose. The spores are of one kind, many and yellowish.

The prothalli are tuberous, lack chlorophyll, and develop below the soil surface.

Two genera are represented in California.

KEY TO THE GENERA

Leaves simple, veins reticulate; sporangia borne in
  rows on elongated spike .... *Ophioglossum*
Leaves mostly lobed or divided into a number of
  segments, veins branching with two divisions of
  same size; sporangia usually in clusters on
  elongated spike or panicle .... *Botrychium*

ADDER'S TONGUE FERNS (*Ophioglossum*)

The leaves are simple, ovate to lanceolate, and have reticulate venation. The sporophyll of the plant is spike-like, bearing its sporangia in two rows.

The generic name is from the Greek *ophis*, snake, and *glossa*, tongue; the narrow fertile spike suggests the name "adder's tongue."

KEY TO THE SPECIES

Blade lanceolate, main veins from 3 to 7,
  ½ of an inch to 1 inch long .... *O. californicum*
Blade ovate to elliptic, main veins from 8 to 20,
  1 inch to 4 inches long .... *O. vulgatum*

[38]

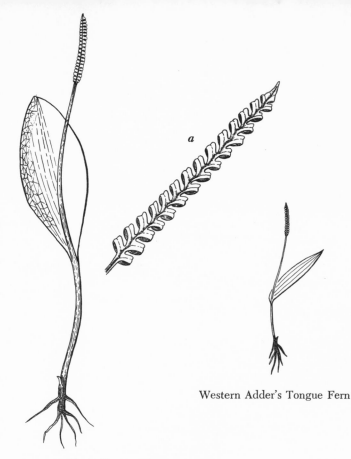

Western Adder's Tongue Fern

Adder's Tongue Fern
a, enlarged view of spore-bearing spike.

Western Adder's Tongue Fern (*Ophioglossum californicum*)
Structural features: Plant small, 1 to 3 inches high. Leaf blade
½ of an inch to 1 inch long, lanceolate, rounded to acute at tip,
with 3 to 7 main veins. Sporophyll ¼ to ¾ of an inch long,
with approximately 15 sporangia on each side.
Habitat: Flats or slopes wet in rainy season.
Range: Amador (near Ione); Monterey and San Diego counties.
December–April.

Adder's Tongue Fern (*Ophioglossum vulgatum*)

Structural features: Plant larger than species above, 9 to 14 inches high. Leaf blade 1 to 4 inches long, ovate to elliptic, sometimes oblanceolate, rounded at tip, usually sessile, with 8 to 20 main veins. Sporophyll ¾ of an inch to 2 inches long, long-stalked, exceeding leaf blade, with 10 to 50 sporangia on each side.

Habitat: Restricted to swampy areas.

Range: Siskiyou County. July.

Early plant naturalists believe that this species had certain curative properties. The juice extracted from the leaves, mixed with the juice from various species of horsetails, was used to curb vomiting and nose-bleeds and to treat ulcers.

MOONWORTS OR GRAPE FERNS (*Botrychium*)

The leaves are mostly lobed or divided into a number of smaller segments (leaflets), which have dichotomous venation. The sporophylls are usually small, face the leaf, and contain clusters of leathery sporangia.

The generic name is from the Greek *botrys,* a cluster, referring to the grapelike arrangement of the sporangia.

KEY TO THE SPECIES

Leaves well above soil surface, arising from above
  middle of stem .... *B. lunaria*
Leaves arising from near base of plant.
    Blades from ½ of an inch to 1¼ inches long, pinnately
      divided; buds without surface hairs .... *B. simplex*
    Blades from 4 to 16 inches long, ternately
      decompound; buds hairy .... *B. silaifolium*

Moonwort (*Botrychium lunaria*)

Structural features: Plant usually stout, fleshy, 1½ to 12 inches high. Leaves arising from middle of stem, 1 inch to 5½ inches long, pinnately divided. Leaflets fan-shaped, crenate or entire at margins, closely overlapping. Sporophyll ½ of an inch to 4¾ inches long, often longer than main leaf, once or twice pinnate.

Habitat: Open places, meadows, slopes, and banks, at about 7,000 feet.

Range: San Bernardino Mountains; Modoc County. Species not abundant.

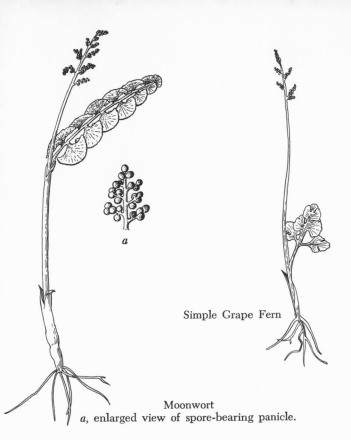

Simple Grape Fern

Moonwort
a, enlarged view of spore-bearing panicle.

The usual form in California is the variety *minganense*. The characteristics are as above, except that the leaflets are farther apart and the blade is more than twice as long as wide. This variety grows well above 7,000 feet in the San Bernardino Mountains, eastern section of the San Gabriel Mountains, the Sierra Nevada of Tulare County, and the Warner Mountains.

Early plant records seem to indicate that juice extracts from this plant were used to stop bleeding and vomiting, and also for the treatment of bruises. They may have been used to concoct balsams for healing internal wounds.

California Grape Fern

Simple Grape Fern (*Botrychium simplex*)

Structural features: Plant slender, 2 to 6 inches high. Leaves arising near base of plant, ovate in shape, ½ of an inch to 1¼ inches long, pinnately divided. Leaflets obovate in shape, rounded at tip, crenulate or lobed at margin. Sporophyll 1¼ to 5 inches long, simple or divided into a number of segments, sometimes reduced to a few sporangia. Buds without surface hairs.

Habitat: Open places, grassy meadows, and damp places, from about 5,000 to 11,500 feet.

Range: San Bernardino Mountains; Sierra Nevada from Tulare County to Tuolumne County; Siskiyou County.

California Grape Fern

(*Botrychium silaifolium* var. *californicum*)
(In Munz, *B. multifidum* ssp. *silaifolium*)

Structural features: Plant stout, fleshy, rather large, 4 to 18 inches high. Leaves solitary or sometimes rounded-triangular or irregular in outline, ternately decompound, 1¼ to 10 inches long, 2¾ to 14 inches wide, thick and fleshy leathery to membranous; ultimate segments many, obovate, obtuse, crenulate or lobed at margins. Sporophyll 6 to 24 inches long, 2 to 5 times pinnate, long-stalked, stout. Buds hairy over surface.
Habitat: Moist meadows and woodland areas, below 7,000 feet.
Range: Sierra Nevada from Tulare County northward to Modoc, Siskiyou, and Del Norte counties; Coast Ranges from Monterey County northward.

At higher elevations, from 7,500 to 11,000 feet, is the variety *coulteri*, which is similar to the form above except that the ultimate segments of the leaves are crowded and the leafstalk is less than 4 inches long. *Range*: Mt. Pinos, Sierra Nevada, White Mountains, and San Bernardino Mountains.

FERN FAMILY (*Polypodiaceae*)

The members of this family are the most conspicuous and frequently encountered ferns in California. They are extremely diverse in habit, size, structure, and habitat—growing in almost all types of conditions. The mature plants are leafy, herbaceous, perennial, and usually evergreen. The underground stems are usually creeping, sometimes erect and stout, and give rise to the lobed or divided leaves. Typical for this family are the stalked sporangia, which are commonly borne in clusters (sori) on the undersurfaces of the leaves. The sporangia are either with or without a protective membranous covering (indusium) and form spores all of one kind. The sori vary in their shape and position on the leaf surface and are therefore important in identification. (See other characteristic features of this family by referring to the Life Cycle of a Typical Fern.)

The prothalli are typically green, heart-shaped, simple in structure, and commonly develop on the surface of the soil.

KEY TO THE GENERA

SORI FOLLOWING COURSE OF VEINS OR
  COMPLETELY COVERING THEM .... *Pityrogramma*, p. 46
SORI OBLONG, LINEAR TO LUNATE OR HORSESHOE-SHAPED
    Sori large, borne in chainlike rows parallel
      to midrib; $\frac{1}{16}$ to $\frac{1}{4}$ of an inch long .. *Woodwardia*, p. 47
    Sori small, oblique, not as above; less than
      $\frac{1}{16}$ of an inch long.
          Sori straight .... *Asplenium*, p. 48
          Sori curved .... *Athyrium*, p. 49
SORI ROUND TO OVAL
    Indusium lacking .... *Polypodium*, p. 56
    Indusium usually present.
      Blade firm or leathery, usually once-
        divided, but in some species at least
        lower pinnae are pinnately lobed or divided;
        indusium centrally attached ... *Polystichum*, p. 61
      Blade not as above, usually more
        than once-divided throughout; indusium
        not attached as above.

Blade 10 to 25 inches or more long
and 4 to 8 inches wide; leafstalks
stout, $\frac{1}{16}$ to almost $\frac{1}{4}$ of an inch
in diameter; indusium kidney-shaped,
thick, conspicuous .... *Dryopteris*, p. 66
Blade 2 to 9 inches long and
2 to 3$\frac{1}{4}$ inches wide; leafstalks
slender, less than $\frac{1}{16}$ of an inch
in diameter; indusium not as above.
Indusium attached beneath
sporangia and surrounding them, divided
into finger-like segments . *Woodsia*, p. 69
Indusium attached at side, with a broad
base, and hooded ... *Cystopteris*, p. 71

SORI BORNE AT OR NEAR MARGINS OF LEAVES
Leaves of 2 kinds, fertile ones larger and with
narrower segments than sterile ones.
Leaves once-pinnate .... *Struthiopteris*, p. 73
Leaves 2 or 3 times pinnate ... *Cryptogramma*, p. 73
Leaves alike or nearly so.
Underground stems (and blades) with hairs
only; plants large and coarse; leaves from
1 foot to 4 feet long, with light-colored stalks;
indusium double, the inner one small,
concealed .... *Pteridium*, p. 74
Underground stems scaly; plants small; leaves
less than 12 inches long, with mostly brown
or purplish stalks; indusium (if any) single.
Sori clearly evident and not continuous
along margins; leaflets distinctly
palmately veined .... *Adiantum*, p. 76
Sori in a more or less continuous
marginal band; leaflets not as above.
Foliage usually without surface materials,
e.g., hairs, scales, etc. .... *Pellaea*, p. 78
Foliage usually white-powdery,
hairy, or scaly.
Ultimate segments often beadlike;
sori with a proper indusial
covering .... *Cheilanthes*, p. 83
Ultimate segments not beadlike;
sori without a proper indusial
covering, the margins of segments
more or less recurved and covering
sporangia ...... *Notholaena*, p. 87

[45]

Golden Back Fern
a, pinnule with sporangia.

GOLDEN BACK FERNS (*Pityrogramma*)

These small to medium-sized ferns prefer dryish banks and rocky outcrops. The underground stems are short, erect, and covered with brownish scales. The leaves are erect to drooping, clustered, uniform, 1 to 3 times pinnately divided, the underside covered with a whitish to yellowish powder. The sporangia follow the course of the veins or completely cover them. The indusium is lacking.

The generic name is from the Greek *pityron*, bran, and *grammae*, letter or line, because of the dandruff-like or scaly linear appearance of the sporangia.

We have only one species; however, several different growth forms have been collected and have been given variety status.

Golden Back Fern (*Pityrogramma triangularis*)

Structural features: Leaves many, 6 to 16 inches long. Leafstalks stout, frequently twice as long as the blades, dark brown, shiny. Blades somewhat triangular or pentangular, once- or twice- pinnate compound, leathery, green, glabrous above. Pinnules oblong, nearly entire to pinnately lobed.

Habitat: Rocky shaded slopes, from 100 to 5,000 feet.

Range: Most of cismontane California from northern Lower California and northward.

GIANT CHAIN FERNS (*Woodwardia*)

These coarse, very large ferns of shady places have short, creeping to erect, underground stems. The leaves are several to many, tufted, and pinnate. The sori are large (1/16 to 1/4 of an inch long), oblong-linear, and borne in chainlike rows parallel to the midrib of the segments. The indusium is elongated, convex, and initially covers all the sporangia.

This fern was named in honor of T. J. Woodward, a noted English botanist.

We have only one species.

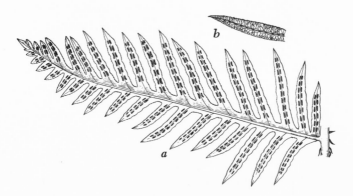

Giant Chain Fern
a, pinna with sori; *b*, pinnule showing venation.

Giant Chain Fern (*Woodwardia fimbriata*)
Structural features: Underground stems woody, with brownish scales. Leaves long-stalked, almost erect, in a circle, 3 to 6 or 9 feet long, oblong-ovate in outline. Pinnae linear-oblong to ovate, deeply pinnatifid, 3 to 12 inches long; their segments lanceolate and spinulose-serrate (toothed).
Habitat: Wet places in mountains, in canyons, and along foot hills where there is year-round seepage, from 100 to 5,000 (or 8,000) feet.
Range: Widely distributed throughout state, but not very common in Sierra Nevada.

This plant is very attractive and is frequently grown in home gardens for decorative purposes.

SPLEENWORTS (*Asplenium*)

Spleenwort is a medium-sized fern of forests and rocky places, with creeping to erect underground stems and rigid scales. The leaves are tufted, pinnate, and herbaceous to slightly leathery. The sori are small (less than 1/16 of an inch long), oblong to narrowly-linear, and oblique to the midrib of the segments. The indusium is fixed lengthwise by one edge.

The generic name is from Greek *a*, without, and *splen*, spleen, referring to several species once used in treating disorders of the spleen.

Three species of *Asplenium* are known for the state. Two of these are not very abundant: *A. viride* is found north of Sierra City, Sierra County; *A. septentrionale*, in Tulare County. The common species is *A. vespertinum*.

Western Spleenwort (*Asplenium vespertinum*)
Structural features: Leaves evergreen, wide-spreading, 2 to 11 inches long, once-pinnate, with short, stout, purplish-brown, shiny stalks. Pinnae 20 to 30 pairs, subsessile, oblong or linear-oblong, with crenate margins. Pinna with 4 to 6 sori.
Habitat: Moist crevices, beneath overhanging rocky cliffs, below 3,000 feet.
Range: Santa Monica Mountains (Ventura County) southward to San Diego County.

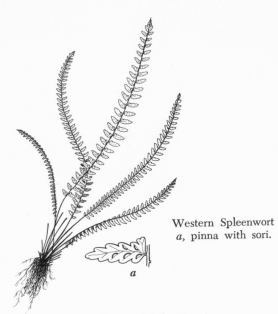

Western Spleenwort
a, pinna with sori.

a

Lady Ferns (*Athyrium*)

Lady ferns are usually medium-sized, upright, graceful plants, growing in moist, shady places. The underground stems are creeping to semi-erect, short-branched, and slightly scaly. The leaves are clustered, erect-spreading, usually large, .1 to 3 times pinnate, and with short to long stalks. The sori are numerous, oblong or linear, often curved and develop obliquely on the midrib. The indusium, if present, is similar in shape to the sorus and is often curved or hooked at the tip.

The generic name is from the Greek *a*, without, and *thurium*, referring to a large oblong shield, the application of which is uncertain.

Lady Fern (*Athyrium filix-femina* var. *californicum*)
Structural features: Leaves erect-arching, lanceolate, slenderly tapering at base and tip, up to 6½ feet long, mostly 6 to 10 inches wide, 2 or 3 times pinnate. Pinnae divided, with lower

[49]

AQUATIC COMMUNITY

Simple Grape Fern

Hairy Pepperwort

Bolander's Quillwort

Giant Horsetail

Duckweed Fern

Common Scouring-Rush

Western Quillwort

ROCKY OUTCROP

Sierra Cliff-Brake

Brewer's Cliff-Brake

American Rock Brake

Hansen's Selaginella

Coville's Lip Fern

Lace Fern

REDWOOD FOREST

Sword Fern

Giant Chain Fern

Western Bracken

Lady Fern

Common Maiden-Hair

Golden Back Fer

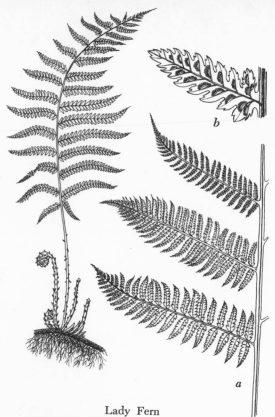

Lady Fern
a, portion of blade; *b*, pinnule with sori.

ones stalked and upper sessile, otherwise similar to leaf. Larger pinnules crenate to pinnate. Sori oblong to horseshoe-shaped. Indusium ciliated, sometimes toothed at margins.

Habitat: Usually moist, shady places along lowlands in mountains, mostly from 3,500 to 8,000 feet.

Range: Along coast, Santa Barbara County to Humboldt County, San Jacinto and San Bernardino Mountains; Sierra Nevada northward to Modoc and Siskiyou counties.

This species is widely distributed both in North and South America. Because of its wide distribution, many variable growth forms have been collected. In Cali-

[56]

fornia we have at least two other variable forms: *A. felix-femina* var. *sitchense* and *A. alpestre* var. *americanum*.

Many people consider that this plant has certain chemical properties for medicinal use. The underground stems, pulverized to a powder, have been used to drive worms out of the intestinal system. This use has not been medically recognized.

POLYPODY FERNS (*Polypodium*)

These small shade-loving ferns have various habitats, but are commonly found growing among rocks or on trees. The underground stems are slender, creeping, and scaly. The leaves are somewhat scattered, all about the same size, herbaceous to leathery, with simple to pinnate blades. The sori are without an indusium, round to oval, and nearly terminal on the tips of the smaller veins.

The generic name is from the Greek *polys*, many, and *podi*, foot, signifying the numerous knoblike branches of the underground stem.

KEY TO THE SPECIES

Blades leathery; sori large (a little over ⅛ of
an inch across); underground stems covered or
    whitened with a bloom. Coastal .... *P. scouleri*
Blades herbaceous; sori small (less than ⅛ of
an inch across); stems not as above.
    Segments pointed, with sawlike teeth at
    margins; blades mostly 6 to 20 inches long.
    Below 4,000 feet.
        Blades usually lanceolate, the segments
        linear-attenuate, mostly spreading, and
        with mostly translucent veins; underground
        stem with a licorice taste .... *P. glycyrrhiza*
        Blades oblong to deltoid, the segments
        linear-oblong, generally close together, and with
        mostly opaque veins. Common ... *P. californicum*
    Segments rounded at tips, with entire
    or crenate margins; blades mostly less
    than 6 inches long. From 5,000 to 8,500
    feet. Rare .... *P. vulgare*

[57]

Leather-Leaf (*Polypodium scouleri*)

Structural features: Underground stems ¼ to ⅜ of an inch thick, covered or whitened with a bloom, without a licorice taste, somewhat scaly, with the scales brown, naked in age. Leaves few, leathery, mostly 6 to 28 inches long, with stout, naked stalks, shorter than blades. Blades deltoid-ovate, 4 to 16 inches long, divided into 2 to 14 pairs of linear to narrowly oblong, obtuse, spreading, crenate segments. Midribs scaly beneath, the smaller veins joined in a simple series of reticulations. Sori large, roundish, without indusium, crowded against midrib, mostly confined to upper segments.

Habitat: Mossy tree trunks, cliffs, and rocky ledges, below 1,500 feet.

Range: Along coast from Santa Cruz County northward to Del Norte County; Santa Cruz Island.

Licorice Fern (*Polypodium glycyrrhiza*)

Structural features: Underground stems ⅛ to ³⁄₁₆ of an inch thick, often compressed, licorice-tasting, with deciduous, deltoid-ovate, rusty brown scales. Leaves few, herbaceous, not evergreen, 8 to 28 inches long, with stalks usually much shorter than blades. Blades usually lanceolate, with segments linear-attenuate, sawlike at margins, generally spreading, with mostly 3-forked translucent veins. Sori oval, without indusium, slightly inframedial.

Habitat: Rocks, logs, and mossy tree trunks, etc., below 2,000 ft.

Range: Along coast from Monterey County and northward.

From earliest times, *P. glycyrrhiza* has attracted collectors of medicinal herbs on account of its licorice-tasting underground stem. Many of the early settlers used these stems to flavor tobacco.

California Polypody (*Polypodium californicum*)

Structural features: Characteristics like those of Licorice Fern, except that underground stems are ³⁄₁₆ to ⅜ of an inch thick; their scales ⅛ to ³⁄₁₆ of an inch long; leaf blades oblong to deltoid; their segments linear-oblong, generally close together; their veins dark, opaque, 3 to 5 times forked.

Habitat: Common on moist, rocky soils and in rock crevices, below 4,000 feet.

Range: Cismontane southern California northward through the Coast Ranges and Sierra Nevada foothills to Humboldt and Butte counties; Santa Barbara Islands.

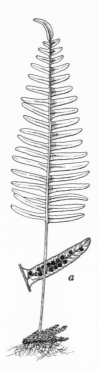

Leather-Leaf
a, pinna with sori.

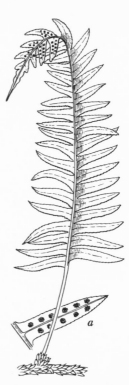

Licorice Fern
a, pinna with sori.

Western Polypody (*Polypodium vulgare* var. *columbianum*) This variety is not widely distributed in California. It has been reported in two areas: the San Jacinto Mountains and the San Bernardino Mountains. It prefers cliffs and rocky slopes, from 5,000 to 8,500 feet. It is similar to the 2 species above, except that the blades are mostly less than 6 inches long, their segments entire or crenate at the margins and rounded at the tips.

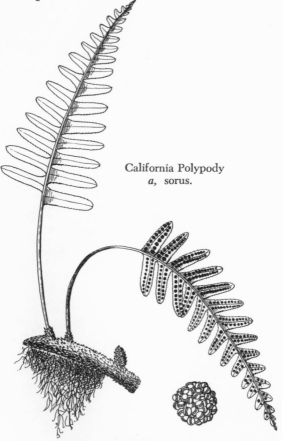

California Polypody
a, sorus.

CHRISTMAS FERNS (*Polystichum*)

These ferns are frequently collected on rocky materials, particularly in damp, wooded places. They are generally stout, large to rather small, and have short-creeping to decumbent, scaly underground stems. The leaves are several, clustered, rigidly ascending or recurved, evergreen, firm or leathery, more or less scaly, and usually once-pinnate, or in some species at least the lower pinnae are again pinnately lobed or divided. The pinnae usually have serrate margins with sharp to spinelike teeth, and the base with a strong lobe at least on the upper side. The sori are round, with the indusium attached at the middle.

The generic name is from the Greek *polys*, many, and *stichos*, row, indicating that the sori in a few species develop in several rows.

Key to the Species

Blades once-divided, the segments (pinnae) variously
  serrate or incised, never lobed or divided.
  Leafstalks ⅜ of an inch to 2⅜ inches long; pinnae mostly
    oblong-lanceolate (lower ones deltoid), serrate-
    dentate, the teeth spreading. Plumas and Siskiyou
    counties .... *P. lonchitis*
  Leafstalks 2 to 20 inches long; pinnae narrowly
    lanceolate, incised to biserrate, the teeth incurved
    Widely distributed .... *P. munitum*
Blades with at least lower segments
  (pinnae) pinnately lobed or divided.
  Teeth of segments not spine-tipped. Siskiyou
    and Trinity counties .... *P. lemmonii*
  Teeth of segments spine-tipped.
    Pinnae lobed or divided, but not completely
      pinnate.
      Older pinnae mostly with 6 lobes on a side,
        more or less deltoid. Pine belt .. *P. scopulinum*
      Older pinnae mostly with 12 or more lobes
        on a side, broadly oblong. Usually below
        pine belt .... *P. californicum*
    Pinnae nearly or quite pinnate, the lowest
      pinnules again cleft or divided .... *P. dudleyi*

Holly Fern (*Polystichum lonchitis*)

Structural features: Underground stems large, woody, with rusty-brown scales. Leaves narrow in outline (tapering toward apex and base), 6 to 24 inches long, with short ($\frac{3}{8}$ of an inch to 2⅝ inches long), scaly stalks. Pinnae numerous, close, mostly oblong-lanceolate (lower ones deltoid), serrate-dentate, with spreading teeth. Sori large, contiguous, usually in 2 rows, nearly in middle of pinae. Indusia entire or nearly so.

Habitat: Rocky, shaded places from 5,000 to 7,000 feet.

Range: Plumas and Siskiyou counties.

This is one of the smaller species of the genus. It has frequently been used to landscape rock gardens.

Sword Fern (*Polystichum munitum*)

Structural features: Underground stems similar to above, except that scales are reddish-brown. Leaves many, sometimes 75 to 100, rigidly ascending in a crown, 1¼ to 5 feet long. Leafstalks 2 to 20 inches long, together with midrib of blade (rachis), conspicuously scaly with bright glossy brown, often dark-centered scales. Pinnae narrow-lanceolate, incised to biserrate, with incurved teeth. Sori submarginal, large, occasionally in 3 rows. Indusia papillose-dentate to long-ciliate.

Habitat: Widely distributed in damp woods generally below 2,500 feet.

Range: Along coast from Monterey County northward to Del Norte County; Santa Cruz Island.

This species shows variability in several characteristics, and for this reason a variety (var. *imbricans*) has been established by several workers. The plant is quite common and is distributed from San Diego County northward to British Columbia.

The Sword Fern is frequently used in floral arrangements such as wreaths.

Shasta Fern (*Polystichum lemmonii*)

This species is not widely distributed. It has been reported from the mountains of Siskiyou and Trinity counties. It seems to prefer moist granitic soils among loose rocks, from about 4,500 to 6,500 feet. It differs from the 2 species above in having the leaf blades with at least the lower segments pinnately lobed or divided; from the other species in having the teeth of the segments not spine-tipped.

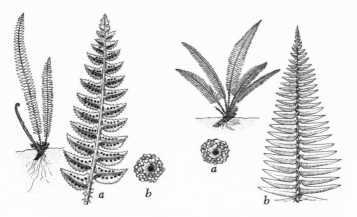

Holly Fern
a, portion of frond; *b*, sorus.

Sword Fern
a, sorus; *b*, portion of blade.

Eaton's Shield Fern (*Polystichum scopulinum*)

Structural features: Underground stems stout, erect or decumbent, with light brown scales. Leaves few, 6 to 10, erect-spreading, 6 to 16 inches long, linear to narrowly oblong-lanceolate, with stout, grooved, densely scaly (at least near base) stalks. Pinnae many, more or less deltoid, lobed or divided near base, with at least 6 lobes on a side, the lobes and teeth spine-tipped. Indusia thin, erose-dentate.

Habitat: Dry cliffs and rocky places, from 5,000 to 10,500 feet. Pine belt.

Range: Northern part of Sierra Nevada to Glenn, Trinity, and Siskiyou counties; San Gabriel and San Bernardino Mountains.

California Shield Fern (*Polystichum californicum*)

Structural features: Underground stems stout, suberect, with large dark brown scales. Leaves ascending, 12 to 44 inches long, with stalks 1½ to 14 inches long, scaly at base. Blades linear-oblong to narrowly linear-lanceolate, with rachis minutely scaly. Pinnae many, linear from broader base, lobed or divided near base, with at least 12 lobes or more on a side, the lobes and teeth spine-tipped. Indusia large, erose-ciliate.

Habitat: Creek banks and canyons, below 1,000 feet.

Range: Along coast from Mendocino County southward to Santa Cruz and Santa Clara counties.

[63]

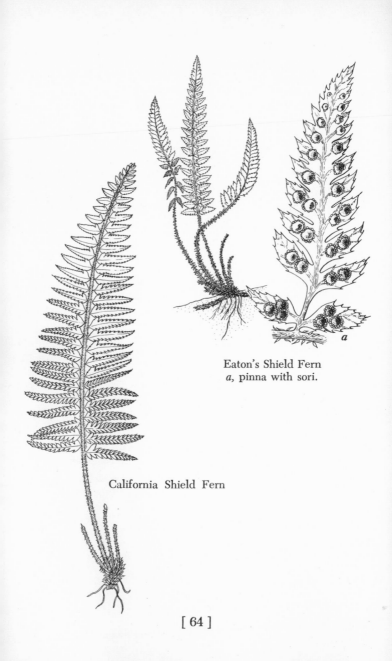

Eaton's Shield Fern
a, pinna with sori.

California Shield Fern

[64]

Dudley's Shield Fern (*Polystichum dudleyi*)

Structural features: Resembles *P. californicum*, except that the pinnae are nearly or quite pinnate; lowest pinnules again cleft or divided; scales denticulate-ciliate; indusia delicate, long-ciliate.

Habitat: Rocky slopes in canyons, below 1,000 feet.

Range: Along coast from northern part of San Luis Obispo County northward to Marin County.

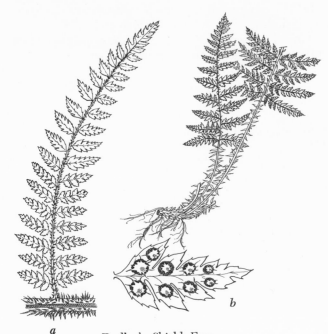

Dudley's Shield Fern
a, pinna; *b*, pinnule with sori.

Wood Ferns (*Dryopteris*)

Wood Ferns are mostly woodland inhabitants. They are large, coarse plants with a slender, short, and stout, wide-creeping, copiously scaly underground stem. The leaves are few to many, herbaceous, often green all winter, borne singly or in a crown, 1 to 3 times pinnate, and their stalks are stout and commonly scaly. The blades are 10 to 20 inches or more long and 4 to 8 inches wide. The sori are round, with the indusium (if present) conspicuous, kidney-shaped, thick, and attached by the inner end of the sinus.

The generic name is from the Greek *drys*, oak, and *pteris*, fern, probably indicating the habitat of some of the species.

Key to the Species

Blades essentially 3 times pinnate .... *D. dilatata*
Blades once or twice pinnate.
    Pinnae without stalks, oblong-lanceolate, the
        lower basal pinnule generally with a semi-
        cordate base and usually overlying main
        rachis; veinlets all ending in salient spine-
        like teeth. Widely distributed .... *D. arguta*
    Pinnae mostly short-stalked, deltoid-lanceolate,
        the basal pinnule not semicordate but sym-
        metrical and apart from main rachis;
        veinlets generally ending in curved teeth.
        San Bernardino Mountains .... *D. filix-mas*

Spreading Wood Fern (*Dryopteris dilatata*)

Structural features: Underground stems woody, creeping or ascending, with brownish scales. Leaves borne in a crown, spreading, 12 to 40 inches long; their stalks stout, dark, covered with brownish, often darker-centered scales. Blades triangular to ovate or broadly oblong, 3 times pinnate or nearly so. Pinnae lanceolate to oblong or basal ones somewhat deltoid. Pinnules lanceolate to oblong, their segments mucronate-dentate at margins. Sori mostly subterminal. Indusia glabrous or sparsely glandular.

Habitat: Shady places in dense woods on decaying logs, below 1,500 feet.

Range: Along coast from San Mateo County northward to Del Norte County.

Spreading Wood Fern
a, portion of pinna with sori.

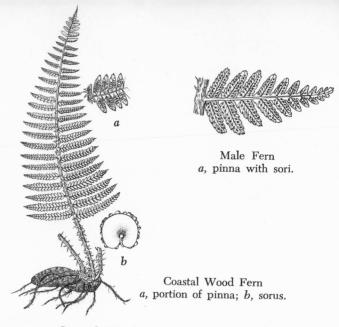

Male Fern
a, pinna with sori.

Coastal Wood Fern
a, portion of pinna; *b,* sorus.

Coastal Wood Fern (*Dryopteris arguta*)

Structural features: Underground stems stout, woody, short-creeping, with numerous bright chestnut scales. Leaves several, close, erect, 12 to 32 inches long; their stalks stout, shorter than blades, scaly. Blades ovate-lanceolate to oblong, acuminate, slightly leathery, twice-pinnate or nearly so. Pinnules oblong-lanceolate, rounded-obtuse, serrate to incise, with mostly spine like teeth. Sori large, close, in two rows. Indusia stiff, with a deep, narrow sinus, the margins minutely glandular.

Habitat: Half-moist stony woods or half-shady slopes in mountains, mostly below 5,000 feet.

Range: Widely distributed; cismontane California from San Diego County and northward.

This species is occasionally grown in gardens.

Male Fern (*Dryopteris filix-mas*)

Structural features: Leaves 12 to 40 inches long; their stalks scaly. Blades similar to *D. arguta,* except that pinnae are short-stalked; their pinnules not subcordate; marginal teeth curved, not spinelike.

Habitat: Rocky woods, at about 8,000 feet.

Range: Reported from a single collection, made in 1882, from Holcomb Valley, San Bernardino Mountains.

The underground stem from this species is processed and frequently sold for worm medicine. It is commonly marketed under the older generic name *Aspidium*. A small amount of this substance taken internally is toxic to flatworms in the digestive tract and drives them out. Some investigators indicate that this material under certain circumstances may be poisonous.

ADDITIONAL SPECIES

Sierra Water Fern (*D. oregana;* in Munz, *Lastrea oregana*). Sierra Nevada from Madera County northward to Plumas County; scattered localities in Trinity and Siskiyou counties.

Downy Wood Fern (*D. feei;* in Munz, *Lastrea augescens*). Wet shaded canyons in parts of southern coastal mountains (Santa Barbara and Los Angeles counties).

Woodsia

These small ferns commonly grow in rocky places. The underground stems are short-creeping or ascending, densely tufted, and clothed with broad, thin scales. The leaves are clustered, numerous, mostly small, linear to lanceolate-ovate, once-or twice-pinnate, and glabrous, variously hairy, or scaly. The sori are round and seated on the back of the free veins. The indusia are under the rounded sori, parting into slender, finger-like divisions often concealed by the sporangia.

The genus is named for Joseph Woods (1776-1864), an English architect and botanist.

KEY TO THE SPECIES

Blades without hairs, bright green; indusial divisions
    linear and beaded .... W. *oregana*
Blades finely glandular to distinctly pubescent.
    dull green; indusial divisions broader at base .. W. *scopulina*

Both species are fragile and delicate, and this makes them popular for use in rock gardens.

Oregon Woodsia (*Woodsia oregana*)

Structural features: Leaves 2 to 10 inches long; their stalks straw-colored, brownish at base. Blades oblong-lanceolate to linear, without hairs, bright green, with 6 to 12 pairs of mostly deltoid-oblong pinnae with toothed lobes. Indusial divisions linear and beaded.

Habitat: Dry rocky places, frequently in limestone, from 4,000 to 11,000 feet.

Range: Riverside County (Santa Rosa and San Jacinto Mountains) through higher mountains of Mojave Desert; Sierra Nevada to Modoc County.

Rocky Mountain Woodsia (*Woodsia scopulina*)

Structural features: Similar to *W. oregana,* but with blades finely glandular to distinctly pubescent, dull green; indusial divisions with broad bases.

Habitat: Exposed rocks, from 4,000 to 12,000 feet.

Range: San Bernardino Mountains; Sierra Nevada; White Mountains.

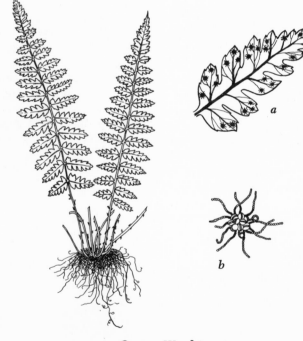

Oregon Woodsia
a, pinna; *b,* sorus.

Brittle Bladder Fern
a, pinnule with sori.

BLADDER FERNS (*Cystopteris*)

Bladder Ferns are non-evergreen, rather delicate
plants that grow in various habitats. The leaves are
clustered, herbaceous, small, and pinnate; those with
sori usually have smaller blades and longer stalks than
those without sori. The sori are numerous, roundish,
separate, and develop on the back of a straight fork
of a vein. The indusium is hoodlike, attached at the
inner side of the broad base.

The generic name is from the Greek *cystis*, a bladder, and *pteris*, a fern, referring to the inflated indusium.

We have only one species.

Brittle Bladder Fern (*Cystopteris fragilis*)

Structural features: Underground stems with thin, ovate scales
near tip. Leaves few to several, with slender, brittle stalks.
Blades oblong-lanceolate, twice pinnate, with pinnules incised
or pinnatifid, decurrent along margined or winged rachis, toothed. Indusia deeply convex.

Habitat: Moist soil of ledges, rocky slopes, and woods; mostly
above 3,500 feet in southern California and lower in northern
California, ascending to 12,000 feet.

Range: Frequent in montane and cismontane California; scattered localities in desert mountains.

[71]

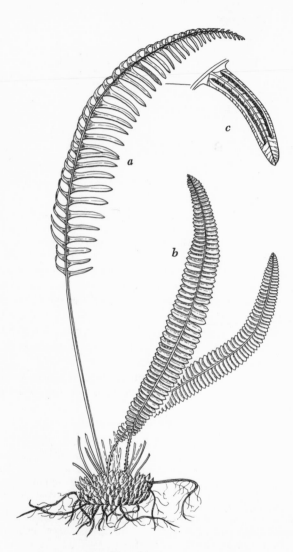

Deer Fern
a, fertile leaf; *b*, sterile leaf; *c*, pinna with sori.

DEER FERNS (*Struthiopteris*)

Deer Ferns are medium-sized, mainly forest inhabitants. The underground stems are mostly stout, erect to ascending, and densely covered with narrow dark scales. The leaves are in tufts, leathery, usually once-pinnate, of 2 kinds: the sterile ones many, spreading or ascending, and with broad pinnae; the fertile ones few, longstalked, and with narrow pinnae. The sori are elongate-linear, in a continuous row, and parallel to the midrib of the fertile pinnae (but nearer to the margin than to the midrib). The indusia are elongate and marginal or near the margins.

The generic name is from the Greek *struthios*, ostrich, and *pteris*, fern, possibly referring to the resemblance of the leaves to ostrich plumes.

Only one species has been collected in the United States.

Deer Fern (*Struthiopteris spicant*)
(In Munz, *Blechnum spicant*)

Structural features: Leaves erect, without hairs. Sterile leaves evergreen, on short stalks, 6 to 40 inches long, lance-linear, with pinnae numerous, crowded, broad. Fertile leaves taller than sterile ones, more erect, long-stalked, with pinnae distant, mostly linear from a broad base, nearly covered by sporangia.
Habitat: Damp, sheltered places.
Range: Along coast from Santa Cruz County northward to Del Norte County.

ROCK BRAKES (*Cryptogramma*)

Rock Brakes are rather small ferns of rocky places, mostly at high altitudes. The underground stems are stout, suberect or somewhat creeping, and covered with thin brownish scales. The leaves are usually many, densely tufted, 2 or 3 times pinnate, without hairs, and of 2 kinds: the fertile ones are larger and have narrower segments than the sterile ones. The sori are many and marginal or nearly so at the end of the veins. The indusium is always present, continuous, and formed from the infolded margin of the pinnules.

[73]

The name is from the Greek *cryptos,* hidden, *gramme,* line, referring to the sori being covered by the infolded margin of the pinnules.

We have only one species.

American Rock Brake or Parsley Fern
(*Cryptogramma acrostichoides*)

Structural features: Fertile leaves simple, 4 to 12 inches long, long-stalked; pinnules linear-oblong, with infolded margins. Sterile leaves smaller than fertile ones, short-stalked, ovate to ovate-lanceolate, with narrow-winged rachises; pinnules ovate, crenate or slightly incised.

Habitat: Rock crevices and rocky slopes, from about 5,500 to 12,000 feet.

Range: San Jacinto and San Bernardino Mountains; White Mountains; Sierra Nevada to Modoc and Humboldt counties.

BRAKE OR BRACKEN FERNS (*Pteridium*)

Bracken Ferns are plants of sunny or shaded places, with creeping, freely branched, woody, black, hairy underground stems. The leaves are many, stout, erect or reclining, 4 feet long, 1 to 3 times pinnate, leathery, with light-colored stalks. The sporangia are marginal, linear, continuous, and covered by a double indusium formed from the infolded margin of the pinnules.

The generic name is from the Greek *pteris,* a wing, applied to ferns because of their feathery leaves.

We have only one species representing this genus; it is considered as a variety.

Western Bracken
(*Pteridium aquilinum* var. *lanuginosum*)

Structural features: Leaves deltoid-ovate, not evergreen, usually 3 times pinnate in lower part. Pinnules oblong-lanceolate, variously hairy on lower surfaces, the upper undivided, the lower more or less pinnatifid.

Habitat: Open woods, rock slides or slopes in damp or dry places, up to 10,000 feet.

Range: Widely distributed throughout state.

The young leaves and underground stems have been collected and used in several ways: as a source of food, as a worm medicine, and as an astringent drug. The older leaves were used by early California settlers for thatching summer shelters, and they are presently used by many hunters, fishermen, etc., for making a soft base for sleeping. The older tissues of the plant have been suspected of being poisonous.

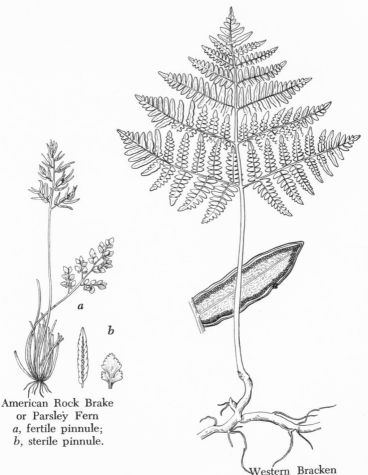

American Rock Brake
or Parsley Fern
a, fertile pinnule;
b, sterile pinnule.

Western Bracken
a, pinnule with sporangia.

MAIDEN-HAIR FERNS (*Adiantum*)

These delicate, graceful ferns grow in moist, shaded places. The underground stems are scaly, long-creeping or short and suberect. The leaves are herbaceous, somewhat bunched, ascending to drooping, and have dark, highly polished stalks. The blades are simple, 2 or 3 times pinnate, mostly broad, and with the pinnules glabrous or variously pubescent and sessile or stalked. The sori are elongate on the underside of reflexed, marginal, indusium-like lobes of the pinnules.

The generic name is from the Greek *a*, not, and *diaine*, to wet, as the foliage sheds rain.

All the species are frequently used to landscape rock gardens.

KEY TO THE SPECIES

Blade divided at base into 2 equal parts, each
   curving part with several pinnae .... *A. pedatum*
Blade not divided at base, bearing pinnae along
   main axis.
      Pinnae with irregular outlines, the outer edge
         deeply lobed or incised .... *A. capillus-veneris*
      Pinnae with regular outlines, the outer edge
         rounded, scarcely lobed ....*A. jordani*

Five-Finger Fern (*Adiantum pedatum* var. *aleuticum*)
Structural features: Underground stems short, thick, with brownish scales. Leaves close, erect 12 to 30 inches long. Blades divided into 2 equal parts, each curving part with 3 to 8 pinnae. Leafstalks stout, dark, scaly at base. Pinnae long-oblong, 3 to 9 inches long; their pinnules short-stalked, mostly oblong, the lower margin entire, the upper incised. Indusia long-oblong.
Habitat: Moist, shaded cliffs or wet, rocky banks throughout state, from sea level to 11,000 feet.
Range: San Bernardino and San Gabriel Mountains through Coast Ranges and Sierra Nevada; Santa Cruz Island.

The Indians used this plant in preparing an ointment for inflammation of the skin.

Common Maiden-Hair (*Adiantum capillus-veneris*)
Structural features: Underground stems creeping, slender, with light brown scales. Leaves often spaced, laxly ascending or pendent, 8 to 28 inches long; their stalks and rachises black and shiny. Pinnae with 3 to 10 pinnules, these wedge-shaped,

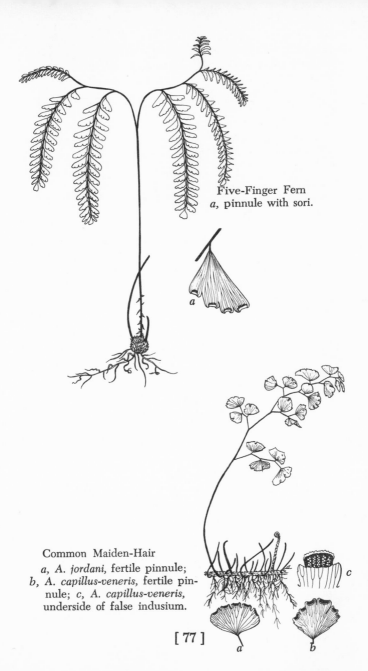

Five-Finger Fern
a, pinnule with sori.

a

Common Maiden-Hair
a, *A. jordani*, fertile pinnule;
b, *A. capillus-veneris*, fertile pinnule; *c*, *A. capillus-veneris*,
underside of false indusium.

c

a   *b*

[77]

deeply lobed or incised at outer edge, with irregular outlines. Indusia transversely oblong, separated.

Habitat: Moist, usually limestone ledges and cliffs or wet, rocky banks, mostly below 4,000 feet.

Range: Widely scattered throughout state, particularly in the southern part.

California Maiden-Hair (*Adiantum jordani*)

Structural features: Resembles species above, but with leafstalks shorter; pinnules broadly fan-shaped, the outer edge rounded, scarcely lobed (or with only 2 to 35 very short lobes in fertile pinnules), with regular outlines; indusia transversely elongated, only slightly interrupted at margins.

Habitat: Moist rocky canyons at low altitudes, mostly below 3,500 feet.

Range: Coast Ranges from San Diego County northward to Oregon border; occasionally in Sierra Nevada foothills; islands off southern California coast.

CLIFF-BRAKES (*Pellaea*)

Cliff-Brakes are rather small plants frequently found in rock crevices, often of calcareous nature. They have short or slender, scaly underground stems and erect, somewhat tufted, persistent leaves. The leaves are numerous, uniform in size, without hairs, firm-herbaceous to leathery, 1 to 4 times divided (pinnate), and with wiry, dark-colored, usually shiny stalks. The sporangia are more or less continuous and borne at or near the margins of the segments. The indusia are formed from the reflexed margins of the segments.

The generic name is from the Greek *pellos*, dusky, referring to the appearance of the leafstalks.

KEY TO THE SPECIES

Blades divided (pinnate) only once.
    Segments entire, leathery, without visible
      veins .... *P. bridgesii*
    Segments mostly 2-parted (mitten-shaped), thin,
      with visible veins .... *P. breweri*
Blades 2 to 4 times divided (pinnate).
    Pinnules oval or ovate, 3/16 to 5/16 of an
      inch wide; underground stems slender and
      creeping .... *P. andromedaefolia*

Pinnules linear, mostly less than ⅜₆ of
an inch wide; underground stems stout, not
long-creeping.
Pinnae divided once or partly twice .. *P. mucronata*
Pinnae divided only once.
Pinna axis several times as long
as each segment and with 11 to 17
pinnules .... *P. compacta*
Pinna axis rarely as long as each segment
and with 5 to 11 pinnules .... *P. brachyptera*

Bridges' Cliff-Brake (*Pellaea bridgesii*)

Structural features: Underground stems short-creeping, often massive, with brownish, blackish-striped scales. Leaves numerous, tufted, 4 to 14 inches long, once-pinnate, with copper-colored, glossy stalks; segments entire, leathery, without visible veins.
Habitat: Dry, exposed rocky cliffs, from 6,000 to 11,000 feet.
Range: Sierra Nevada from Tulare County northward to Sierra County.

Brewer's Cliff-Brake (*Pellaea breweri*)

Structural features: Underground stems ascending to decumbent, with twisted, dark-red scales. Leaves very densely tufted, 2 to 8 inches long, once-pinnate, with reddish-brown, glossy stalks; segments mostly 2-parted (mitten-shaped), thin, with visible veins.
Habitat: Exposed rocky cliffs and slopes, from 7,000 to 12,000 feet.
Range: Sierra Nevada from Tulare County northward to Siskiyou County; White, Panamint, and San Bernardino Mountains.

Coffee Fern (*Pellaea andromedaefolia*)

Structural features: Underground stems slender, wide-creeping, with overlapping, narrow, hair-pointed scales. Leaves scattered, evergreen, 6 to 28 inches long, twice pinnate or partly 3 times pinnate, with straw-colored, glaucous stalks; pinnae with 3 to 5 pairs of oval or ovate, short-stalked pinnules.
Habitat: Dry, rocky soils, from 1,000 to 4,000 feet.
Range: Cismontane southern California; Coast Ranges northward to Mendocino County; Sierra Nevada foothills from Tulare County northward to Butte County.

Bird's-Foot Cliff-Brake (*Pellaea mucronata*)

Structural features: Underground stems stout, woody, with closely tufted, chestnut-colored scales. Leaves loosely tufted, evergreen, dry and brittle, 6 to 20 inches long, 2 or 3 times pinnate, mostly fertile, with purplish-brown, rather dull stalks; pinnae divided once or partly twice into 6 to 15 pairs of usually 3-foliolate pinnules.

Habitat: Exposed or partially shady, rocky places in foothills, generally below 6,000 feet.

Range: Throughout California; occasionally on desert.

The younger leaves and stems are occasionally eaten by sheep and other grazing animals. They are, however, poisonous to these animals and frequently cause death. Because of this, they are sometimes called Poison Ferns or Black Ferns.

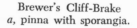

Brewer's Cliff-Brake
a, pinna with sporangia.

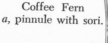

Coffee Fern
a, pinnule with sori.

Bird's-Foot Cliff-Brake
a, pinnule; *b*, segment of pinnule with sporangia.

Desert Cliff-Brake (*Pellaea compacta*)

Structural features: Underground stems stout, woody, with tufted, dark or gray-brown scales. Leaves many, tufted, evergreen, 8 to 14 inches long, twice-pinnate, mostly fertile, with brown stalks; pinnae divided only once into 11 to 17 sessile, linear pinnules. Pinna axis several times as long as each segment.
Habitat: Arid, stony slopes or flats, from 4,500 to 8,500 feet.
Range: Sierra Nevada from Tuolumne County and southward; Panamint, San Gabriel, San Bernardino, and San Jacinto Mountains.

Sierra Cliff-Brake (*Pellaea brachyptera*)

Structural features: Underground stems horizontal, thick, covered with compactly arranged brownish scales. Leaves clustered, 6 to 16 inches long, twice-pinnate, with stout, purplish-brown stalks; pinnae with 5 to 11 pinnules. Pinna axis rarely as long as each segment.

Habitat: Exposed rocky ledges and slopes, from 3,500 to 8,000 feet.

Range: Plumas County to Lake County and northward.

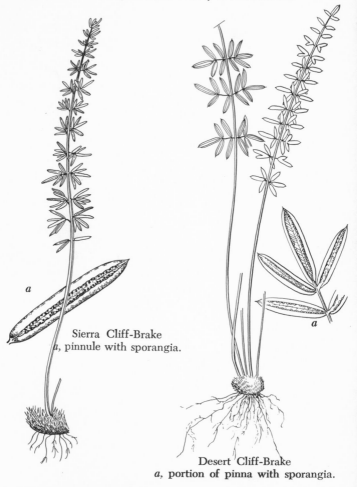

Sierra Cliff-Brake
a, pinnule with sporangia.

Desert Cliff-Brake
a, portion of pinna with sporangia.

Lip Ferns (*Cheilanthes*)

Lip Ferns are small, rock-inhabiting plants usually found growing under extremely dry conditions. The leaves are all of the same size, in clusters, with the blades 1 to 4 times pinnate, or once-pinnate and variously lobed or divided; the ultimate segments are often beadlike and usually hairy or scaly beneath. The sporangia are borne at the margins and have a proper indusial covering.

The generic name is from the Greek *cheilos*, lip, and *anthos*, flower, in reference to the shape of the marginal sori.

Several of the species in this generic group are difficult to distinguish from the species in the genus which follows (Cloak Ferns, *Notholaena*), and for this reason have been shifted from one group to the other by several investigators. Munz and Keck, in their *California Flora*, have taken all the species in both genera and placed them in a single generic category. A number of species in both genera which are not widely distributed are listed in this guide only to show their geographical distribution.

Key to the Common Species

Leaves without hairs, triangular in shape.
    Sori solitary, with separate small, lunate
        indusia .... *C. californica*
    Sori in a continuous row, with a long
        narrow indusium .... *C. siliquosa*
Leaves hairy or scaly, linear to oblong-lanceolate.
    Segments flattish; indusium not continuous
        around margin .... *C. cooperae*
    Segments beadlike; indusium continuous
        around margin.
        Blade densely hairy on underside, not
            scaly; leafstalk not scaly .... *C. gracillima*
        Blade densely scaly on underside;
            leafstalk scaly below ....*C. covillei*

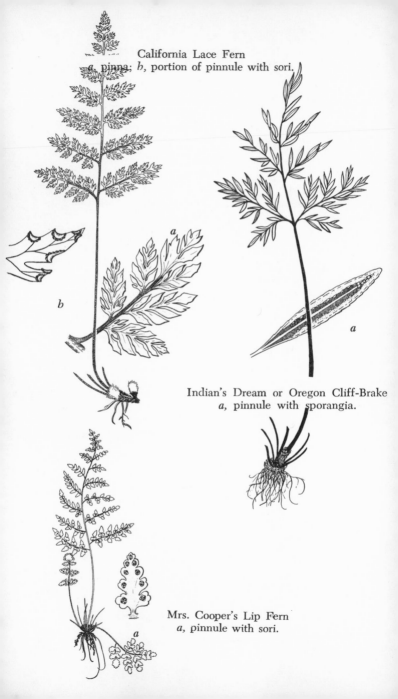

California Lace Fern
a, pinna; *b,* portion of pinnule with sori.

b

a

Indian's Dream or Oregon Cliff-Brake
a, pinnule with sporangia.

a

Mrs. Cooper's Lip Fern
a, pinnule with sori.

a

California Lace Fern (*Cheilanthes californica*)
(In Munz, *Aspidotis californica*)

Structural features: Underground stems short-creeping, covered with chestnut-colored scales. Leaves many, 2⅜ to 14¾ inches long, with shiny, dark brown, densely tufted stalks. Blades triangular in outline, finely 4 times pinnate, without hairs, upper division of rachis with a narrow herbaceous border. Sori solitary, with separate small, lunate indusia.

Habitat: Foothills about rocks and cliffs, generally in shady places, below 2,500 feet.

Range: Very common in Coast Ranges from Humboldt County southward to San Diego County; scattered situations in Sierra Nevada from Butte County southward.

Indian's Dream or Oregon Cliff-Brake
(*Cheilanthes siliquosa;* In Munz, *Onychium densum*)

Structural features: Underground stems with brownish, shiny scales. Leaves many, crowded, 2¾ to 12 inches long, with shiny, chestnut-colored stalks. Blades roundish-ovate, 3 times pinnate below, without hairs, with linear pinnules. Sporangia in a continuous row with a long narrow indusium.

Habitat: Crevices of rocky outcrops, from 6,000 to 8,500 feet and as low as 1,000 feet.

Range: Sierra Nevada from Tulare County northward to Nevada County; Coast Ranges, San Luis Obispo and Kern counties.

Mrs. Cooper's Lip Fern (*Cheilanthes cooperae*)

Structural features: Underground stems short, thick, with tufted scales at top. Leaves numerous, clustered, 2 to 8¾ inches long, oblong-lanceolate, with dark brown stalks. Blades twice-pinnate, covered with gland-tipped hairs, with flattish segments. Sori 1 or 2 to a lobe. Indusium not continuous around margin of segments.

Habitat: Occasionally in deep crevices of rocky cliffs, below 2,000 feet.

Range: Sierra Nevada foothills in following counties: El Dorado, Calaveras, Tuolumne, Mariposa, and northward; Santa Inez Mountains southward in scattered localities to Slover Mountain near Colton.

Lace Fern (*Cheilanthes gracillima*)

Structural features: Underground stems tufted, with numerous short branches, covered with a dense mass of light brown scales. Leaves very numerous, 4 to 10 inches long, linear to lanceolate, with dark brown stalks. Blades twice-pinnate, densely hairy on undersides, with beadlike segments, the margins deeply curved.

Habitat: Rocky outcrops at higher elevations, from 2,500 to 9,000 feet.

Range: Sierra Nevada from Tulare County northward to Lassen County; Coast Ranges from Marin County to Siskiyou County.

Coville's Lip Fern (*Cheilanthes covillei*)

Structural features: Underground stems short, branched, covered with dark brown to blackish scales. Leaves many, 4 to 12 inches long, ovate to ovate-lanceolate, with brown to dark purplish stalks, scaly below. Blades 3 times pinnate, densely covered with scales on undersides, the segments beadlike.

Habitat: Dry, rocky slopes and rock crevices, from 1,500 to 9,000 feet.

Range: Colorado Desert and cismontane southern California; northward in inner Coast Ranges as far as Lake County; Sierra Nevada as far north as El Dorado County; Mojave Desert and northward to Inyo and Lassen counties.

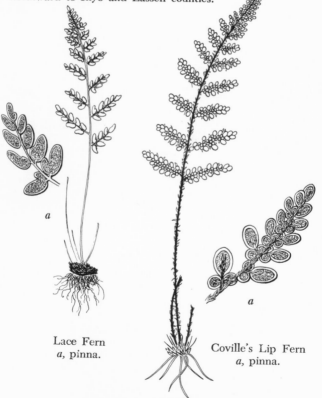

Lace Fern
a, pinna.

Coville's Lip Fern
a, pinna.

ADDITIONAL SPECIES

Viscid Lip Fern (*C. viscida*). Mountainous slopes of western edge of Colorado Desert and Panamint Range.

Slender Lip Fern (*C. feei*). Mountains of eastern part of the Mojave Desert.

Fibrillace Lip Fern (*C. fibrillosa*). San Jacinto Canyon, Riverside County.

Parish Lip Fern (*C. parishii*). Andreas Canyon, eastern base of mountain. San Jacinto, Riverside County.

Cleveland's Lip Fern (*C. clevelandii*). Riverside, San Diego, and Santa Barbara counties; Santa Rosa Islands.

Coastal Lip Fern (*C. intertexta*). Coast Ranges from Santa Clara County northward to Mendocino County; Sierra Nevada from Tulare County northward to Siskiyou County.

Wooton's Lip Fern (*C. wootonii*). Panamint and Inyo-White Mountains, Inyo County; Providence and New York Mountains, San Bernardino County.

CLOAK FERNS (*Notholaena*)

Cloak Ferns are generally similar to Lip Ferns in their growing conditions and structural features, except that the segments are not beadlike, and the sporangia are without a proper indusium: the indusium is formed by the more or less recurved margins of the segments.

The generic name is from the Greek *nothos*, spurious, and *laena*, cloak; the soft woolly hairs beneath form a covering over the sporangia in some members of this group.

KEY TO THE COMMON SPECIES

Leaf surfaces without hairs or nearly so .... *N. jonesii*
Leaf surfaces woolly, especially beneath.
Leafstalks and rachis sparsely covered with
    short, erect, stiff hairs; segments of blades
    stalked and densely woolly. Deserts .... *N. parryi*
Leafstalk and rachis at first densely
    covered with soft woolly hairs; segments of
    blades sessile and very finely woolly.
    Cismontane southern California .... *N. newberryi*

[87]

Jones' Cloak Fern (*Notholaena jonesii*)
(In Munz, *Cheilanthes jonesii*)

Structural features: Underground stems short, oblique, conspicuously covered with light brown scales. Leaves many, 1⅛ to 6⅜ inches long. Leafstalks curved. reddish-brown. Blades twice-pinnate or sometimes more, generally without surface hairs. Sporangia in a broad submarginal band.

Habitat: Rocky places, particularly in limestone, from about 3,500 to 6,000 feet.

Range: Scattered localities in desert ranges of southern California; Mojave Desert northward to White Mountains and westward to Santa Barbara County.

Parry's Cloak Fern (*Notholaena parryi*)
(In Munz, *Cheilanthes parryi*)

Structural features: Underground stems short, densely covered with reddish-brown scales often with a blackish stripe. Leaves many, erect, 2⅜ to 7⅛ inches long. Leafstalks purplish-brown, slender, long, sparsely covered with short, erect, stiff hairs. Blades 3 times pinnate; segments stalked, densely woolly, appearing like pellets of wool. Sporangia few, dark, partly evident with age.

Habitat: Dry, rocky slopes and ledges, usually below 7,000 feet.

Range: Desert regions, from White Mountains southward.

Cotton Fern (*Notholaena newberryi*)
(In Munz, *Cheilanthes newberryi*)

Structural features: Underground stems slender, branched, creeping, covered with blackish-brown, shiny scales. Leaves erect, clustered, 3⅛ to 9½ inches long. Leafstalks wiry, purplish-brown, shiny, at first densely covered with soft woolly hairs. Blades 2 to 3 times pinnate; segments sessile, very finely woolly. Sporangia rather large, blackish, visible with age.

Habitat: Common in crevices of rocky cliffs, below 2,500 feet.

Range: Lower foothills of Coast Ranges, southern California, from Ventura County southward to San Bernardino and San Diego counties; scattered localities on San Clemente Island.

ADDITIONAL SPECIES

Cloak Fern (*N. cochisensis;* in Munz, *Cheilanthes sinuata* var. *cochisensis*). Providence and Clark Mountains, San Bernardino County.

California Cloak Fern (*N. californica;* in Munz, *Aleuritopteris cretacea*). Desert slopes of San Bernardino Mountains in southern California from Victorville to Palm Springs and southward.

[88]

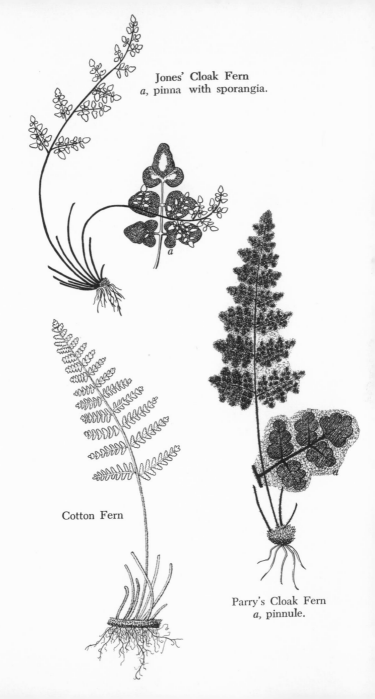

Jones' Cloak Fern
a, pinna with sporangia.

Cotton Fern

Parry's Cloak Fern
a, pinnule.

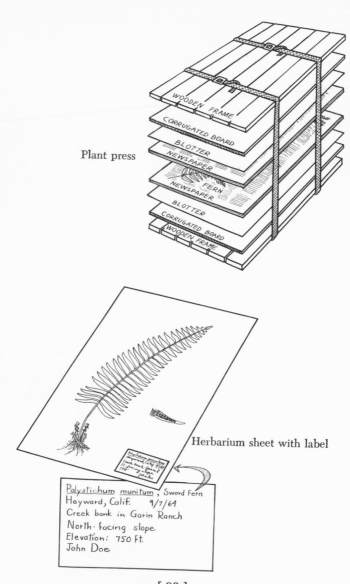

Plant press

WOODEN FRAME
CORRUGATED BOARD
BLOTTER
NEWSPAPER
FERN
NEWSPAPER
BLOTTER
CORRUGATED BOARD
WOODEN FRAME

Herbarium sheet with label

Polystichum munitum , Sword Fern
Hayward, Calif.   9/7/64
Creek bank in Garin Ranch
North-facing slope
Elevation: 750 ft.
John Doe

ACTIVITIES

There are many interesting nature activities which can add greatly to your knowledge and pleasure, besides enabling you to identify fern plants. A few of these are described below.

How to Make a Fern Herbarium

A collection of plants, properly pressed, dried, and mounted on stiff white paper, is known as a herbarium. A herbarium may be small or large, depending upon the number of plants in the collection. This collection may be from a local area, state-wide, or world-wide. Plant collections are valuable because they give a detailed account of the earth's vegetation.

For your fern herbarium try to obtain complete and typical plants in a fertile condition. The underground stem, one or two leaves, and the sori must be available. Avoid collecting the entire underground stem, as this would eradicate the plant. If it is impossible to get a complete plant, it may be necessary to obtain materials from a number of plants. Specimens gathered in the field may be carried for some time without wilting if they are wrapped in wet newspaper and placed in a plastic bag, or they may be pressed directly upon collection.

The pressing and drying process consists of arranging specimens between folded newspaper, placing the folded sheet between two blotters, then a sheet of corrugated pasteboard, then the blotting paper again, and so on. Pressure is applied to remove all the water so that the specimens dry quickly to retain most of their natural color. Their appearance and usefulness are improved by this method. The specimens should be kept in the press for 12 to 48 hours or longer, depending upon the thickness of the plant parts. Wet blotters should be changed several times to hasten the drying process. Field presses for applying pressure on plants may be purchased from a biological supply house for a moderate sum.

If you are planning an extensive herbarium, you might be interested in making your own field press. It should be of tough, durable construction. Strips of wood, about 2 inches wide, may be glued and nailed together to make a standard field press. The usual size is 12 by 18 inches. A pair of straps about one inch wide and 3 to 5 feet long, with buckles, will complete the press. (Instead of straps, you could use strong cord.) Tightening down on the straps will provide the pressure necessary to remove water from the specimens.

Mounting is done on sheets of stiff white paper 11½ by 16½ inches with label-bearing collection data attached at the lower right-hand corner. There are several ways to mount the specimens on the paper. They can be attached by small strips of gummed tape, usually cloth or some other type, or glued with a strong fish glue. The latter provides better mounts and is the standard method used by most professional collectors. This glue can be purchased from a biological supply house. Some collectors prefer to cover the plants with sheets of celluloid or a similar material, providing specimens which are not easily damaged. This method, of course, is expensive and is frequently used when specimens are on exhibit.

The pressed and mounted specimens are not complete without labels showing collection data. These include:

1. Name of plant
2. Place where plant was collected, and date
3. Habitat: type of environment in which plant was growing,
   e. g., meadows, along railroad tracks, etc.

4. Exposure: north-facing slope, etc.
5. Elevation
6. Habit of growth: prostrate, erect, etc.
7. Collector's name

Not all the data above need be included on the label, but it is good practice to record as much of this information as possible, since it may be necessary for identification and reference.

How to Raise Ferns from Spores

Growing ferns from spores is a most rewarding experience and not too difficult. Fern spores, collected from woodland plants or cultivated ferns during the spring or early summer months, will germinate if they are scattered on a porous substrate, like a sandy-leaf mulch mixture, and if proper humidity conditions are maintained.

The simplest method is to fill a shallow container, like a dish or pan, half full of fine gravel or other suitable substrate. Over this material spread a thick layer of a sandy-leaf mulch mixture and then sterilize everything by placing the container in an oven for about half an hour. Allow time for cooling, and then sprinkle a thin layer of fern spores over the surface of the soil mixture. Pour enough water in the container to cover the gravel and invert a large glass container over all the material and stand in a shady to dark, humid place. No further care is needed.

Within three weeks, or possibly longer, depending upon the types of fern spores and environmental conditions, green heart-shaped plants (prothalli) will be seen developing on the surface of the soil mixture. Not all the prothalli will be mature, and one would expect to find these plants in various stages of development. If enough time is allowed for the plants to complete their development you will observe the young fern plants (sporophytes) developing from the notch at the widest part of the prothallus. After a time the prothalli

[93]

will gradually decompose and the young fern plants can then be transplanted into larger containers to allow for their complete development.

Propagation of fern spores

GROWING FERNS FROM THE UNDERGROUND STEMS
Ferns can easily be grown from the underground stems. This method of propagation gives the quickest results, and takes less time than growing ferns from spores. Simply remove a piece of a fern plant, making sure that a part of the underground stem and one or two leaves are present, and transplant this material to a suitable environment. The best time to transplant fern parts from their natural environment is in the late winter or early spring. These materials will remain in good condition for planting for several days, even a week or two, if kept in a cool, shaded spot.

Ferns generally grow best in shady areas where the soil is highly porous and rich in organic material. To make a suitable soil, be sure to add sand and half-decayed leaf mulch. Always make a careful record of the conditions under which native ferns grow, so that you may duplicate these conditions wherever you plan to grow them.

To grow ferns indoors during the entire year, the same conditions are necessary. Remember, however, to place about two inches of gravel or broken pots at the base of the container to provide proper drainage.

ECONOMIC USES

Living ferns are of little beneficial importance to man except as ornamental plants. A few species which form long surface hairs are used for stuffing and packing material. Tropical ferns develop large trunks which are often used for lumber. The ferns of the past are beneficial today because they have contributed, along with other plants, to our enormous coal deposits. A perusal of the literature indicates that many ferns have been collected to determine their value in medicine. Our records indicate that as early as 300 B. C. some of these plants were thought to have medicinal value.

It would be interesting to make a careful study of fern journals, botanical publications, and other references to discover how our native ferns have been used in the past, and whether any are being collected today for other than ornamental purposes.

MAKING PRINTS OF LEAVES

Prints of fern leaves can be made in a number of ways. One way is to place the leaf on drawing paper, stationery, or paper of unusual texture and spray paint over the leaf with a commercial spray-paint can, or a paint-loaded toothbrush over the edge of a ruler on to the specimen. Another way is to place the leaf on a piece of glass or masonite and roll waterproof ink or paint over the leaf and the glass with a rubber roller. This paint-covered leaf is then placed on a sheet of paper. Pressure is applied over the entire surface of the leaf so that the ink is transferred to the paper. These prints may be used for decorative purposes.

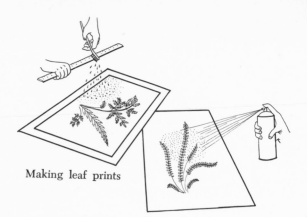

Making leaf prints

DRIED-PICTURE ARRANGEMENT

Several different types of dried fern leaves, processed as described earlier, can be arranged in various ways on paper or wood to produce a dried-picture arrangement. These specimens can be mounted on this backing material, using a good fish glue or some other type of mounting medium. The arrangement can be kept for years if shellac or plastic paint is sprayed over the entire surface.

MAKING PLASTER PLAQUES

Pour wet plaster of Paris to a depth of half an inch or more in a greased container, the size and shape of the desired plaque. Place the leaves in an attractive arrangement in the wet plaster. The leaves can be re-arranged as long as the plaster remains wet. Allow time for the plaster to dry, and then remove the leaves. Paint the plaque with paint or shellac. The plaque is now ready to hang.

BLUEPRINTS AND OZALID PRINTS

Blueprinting fern leaves provides an inexpensive method of seeing differences among fern plants. Blueprint paper can be purchased from a blueprinting or engineering firm; you might even try department or artist's supply stores. This paper is frequently sold in rolls; however, the supplier may make it available to you in various sizes. The standard size is 8½ by 11 inches. Be careful not to expose the paper to light until you are ready to use it. (1) Place a dried fern leaf, processed as described earlier, on a piece of glass of standard size. (2) Cut a piece of heavy cardboard or other backing material the size of the glass and fasten the glass and the backing together along one edge with wide masking tape. (3) Place a sheet of blueprint paper with the sensitive side over the specimen. (4) Turn the glass side up toward the light and expose the paper until it darkens and turns blue. (5) Remove the paper and wash throughly in tap water for about 10 minutes and dry with a sponge or cotton swab. The print will have a blue background, with the fern-leaf design in a clear white. The prints will remain more permanent if they are washed in a weak dichromate solution for approximately 5 minutes. This should be done before washing in tap water. The dichromate solution can be prepared by adding a pinch of potassium dichromate crystals to a glass of water.

Ozalid paper, which comes in a variety of colors, can be purchased from your blueprint supplier, and you can make prints in much the same way. However, this paper is developed by washing in a weak solution of ammonia water rather than in tap water. The prints can also be developed by exposing them to ammonia fumes in a container. When ozalid paper is used, the specimen design will be colored and the background white—just the reverse of the result obtained with blueprint paper.

GLOSSARY

Acuminate — tapering gradually to a point.

Aerial — growing in the air above ground and water, said of parts of plants.

Appressed — lying close and flat against another part, as leaves on stems.

Attenuate — tapering; gradually becoming slender.

Awl-shaped — tapering from a broad base to a sharp point.

Bipinnate — divided once, and then each division divided again, as a leaf blade.

Blade — the flattened, enlarged part of a leaf.

Cilia (singular, **cilium**) — hairlike process from a plant surface.

Cismontane — referring to this side of the mountain, that is, west of the main sierran crests, as opposed to the deserts.

Compound leaf — a leaf with the blade divided into a number of segments.

Cone — an elongated structure composed of modified leaves (sporophylls) on which sporangia develop.

Contiguous — close together, touching one another.

Cordate — heart-shaped.

Corm — an underground stem which stores food.

Crenate — notched; having rounded or blunt teeth, as the margins of leaves.

Cuneate — wedge-shaped.

Deciduous — shedding the leaves at the end of a growing season.

Decompound — divided several times, as some leaves are.

Decumbent — lying on the ground but tending to rise at the end.

Deltoid — triangular in shape.

Dentate — having a toothed margin, said of leaves with sharp, rather coarse teeth pointing outward from the center.

Denticulate — having toothed edges.

Dichotomous — branching or forking with the two divisions nearly equal.

Dimorphic — having two different forms on the same plant, or in the same species.

Dorsal — pertaining to the back or outer surface.

[98]

Dorsoventral — having both dorsal and ventral surfaces.

Elliptic — oblong with rounded ends.

Entire — the margin even, in no way toothed or lobed.

Erose — having an irregularly toothed margin, as if gnawed.

Evergreen — retaining the leaves for more than one growing season.

Fertile — referring to a spore-bearing organ, such as a leaf with sori.

Frond — the leaf of a fern.

Genus (plural, **genera**) — a group of closely related species.

Glabrous — smooth without hair.

Glaucous — covered or whitened with a bloom which is easily rubbed off.

Habitat — the immediate area in which a plant lives.

Herb — a plant which does not develop woody tissues.

Hyaline — thin and almost transparent.

Imbricate — overlapping regularly, like shingles on a roof.

Incised — deeply cut, with indentations of different sizes, as on leaf margins.

Indusium — a membrane developing from leaf tissue and covering the sorus.

Internode — a space between two joints (or two nodes).

Keel — a median longitudinal ridge or process, like the keel of a boat.

Lanceolate — lance-shaped, wide near the base and gradually tapering to the tip.

Linear — long and narrow, with the margins approximately parallel, as a leaf.

Lunate — shaped like a half-moon.

Marginal sori — sori developing at the margins of the segments.

Megasporangium — a spore-case in which the megaspores develop.

Megaspore — a spore which in vascular plants gives rise to a female gametophyte.

Microsporangium — a spore-case in which the microspores develop.

Microspore — a spore which in vascular plants gives rise to a male gametophyte.

Mucronate — tipped with a short, stout projection.

Netted — said of veins branching to form a network.

Node — a point on a stem from which a lateral projection develops, such as a leaf.

Oblanceolate — inversely lanceolate, with the widest portion slightly above the middle and gradually tapering toward the base.

[99]

Oblong — with sides nearly parallel, and the length two or three times the width.

Obovate — inversely ovate, with widest part above the middle, and the narrower end at the base.

Obtuse — with a rounded apex.

Orbicular — rounded or nearly so.

Ovate — similar to the shape of an egg, with the widest part near the base.

Palmate — shaped like a hand, said of a compound leaf with the segments developing from a common center at the end of the leaf stalk.

Panicle — a loose, irregularly branched sporangial cluster.

Papillose — bearing minute bumps or processes.

Peduncle — a leafless stalk bearing strobili (cones) or sporocarps.

Perennial — living several to many seasons, said of plants.

Petiolate — with a leafstalk.

Pinna (plural, **pinnae**) — the first division of a compound fern leaf.

Pinnate — with the leaflets arranged along each side of an axis.

Pinnatifid — with the margin pinnately cleft or parted.

Pinnule — the second division of a compound leaf, and the first division of the pinna.

Pubescent — clothed with soft, downy hair.

Rachis — that portion of the axis of a compound leaf from the lowest pair of pinnae upward.

Reticulate — in the form of a network, said of leaf veins.

Scale — a small, thin, non-green modified leaf.

Serrate — with sawlike teeth, the teeth sharp and pointed forward.

Sessile — without a leafstalk.

Sheath — the base of the blade or leafstalk which encloses the stem.

Simple leaf — a leaf with a single blade, not divided.

Sinus — an indentation between two lobes, as of the indusium.

Sorus — a group of sporangia.

Species — a particular kind of plant (or groups of plants) which maintains its distinctness (or their distinctness) over several generations.

Spicate — spikelike.

Spike — a long sporangial cluster, with the sporangia attached directly to the stalk.

Spinose — covered with spines.

Sporocarp — a receptacle in which sori are enclosed (as in *Marsilea*).

[100]

Sporophyll — a modified leaf upon which spores are borne.

Sterile — lacking spore-producing structures.

Stomate — opening in the epidermis.

Strobilus — a conelike structure consisting of a central axis about which are attached closely packed sporophylls.

Submarginal sorus — a sorus somewhat back but parallel to the margin of a segment.

Subulate — very narrowly triangular.

Ternately compound — referring to a compound fern leaf with three pinnae.

Terrestrial — growing on soil, said of plants.

Tomentose — covered with soft, matted hairs.

Tripinnate — three times pinnate.

Tubercles — small, rounded projections.

Variety — a different form within a species, sometimes referred to as a subspecies.

Venation — the arrangement of veins, as in a blade or pinnae of a leaf.

Ventral — the lower or under surface.

SUGGESTED REFERENCES

The references listed below include sections where detailed information can be obtained for the ferns and related forms.

Abrams, L. *Illustrated Flora of the Pacific States.* Stanford, California: Stanford University Press. Volume 1. 1923.

Frye, T. C. *Ferns of the Northwest.* Portland, Oregon: Metropolitan Press, 1934.

Holt, V. *Key to Wild Flowers, Ferns, Trees, and Shrubs of Northern California.* Palo Alto, California: National Press, 1955.

Jepson, W. L. *Manual of the Flowering Plants of California.* Berkeley, California: University of California Press. 1925.

Munz, P. A. and D. D. Keck. *A California Flora.* Berkeley and Los Angeles: University of California Press. 1959.

Rodin, R. J. *Ferns of the Sierra.* Yosemite Nature Notes. Volume 39, No. 4.

CHECK LIST OF
FERNS AND FERN ALLIES OF CALIFORNIA

HORSETAIL FAMILY (Equisetaceae)

Common Scouring-Rush (*Equisetum hyemale*), p. 17
Braun's Scouring-Rush (*Equisetum laevigatum*), p. 18
California Horsetail (*Equisetum funstoni*), p. 19
Summer Scouring-Rush (*Equisetum kansanum*), p. 20
Common Horsetail (*Equisetum arvense*), p. 21
Giant Horsetail (*Equisetum telmateia* var. *braunii*), p. 21

WATER FERN FAMILY (Salviniaceae)

Duckweed Fern (*Azolla filiculoides*), p. 24

PEPPERWORT FAMILY (Marsileaceae)

Hairy Pepperwort (*Marsilea vestita*), p. 25
Nelson's Pepperwort (*Marsilea oligospora*), p. 26
Pillwort (*Pilularia americana*), p. 27

QUILLWORT FAMILY (Isoetaceae)

Howell's Quillwort (*Isoetes howellii*), p. 29
Nuttall's Quillwort (*Isoetes nuttallii*), p. 29
Orcutt's Quillwort (*Isoetes orcuttii*), p. 29
Western Quillwort (*Isoetes occidentalis*), p. 30
Braun's Quillwort (*Isoetes muricata*), p. 31
Bolander's Quillwort (*Isoetes bolanderi*), p. 31

CLUB-MOSS FAMILY (Lycopodiaceae)

Running Pine (*Lycopodium clavatum*), p. 33
Bog Club-Moss (*Lycopodium inundatum*), p. 32

SPIKE-MOSS FAMILY (Selaginellaceae)

Bushy Selaginella (*Selaginella bigelovii*), p. 34
Douglas' Selaginella (*Selaginella douglasii*), p. 34
Wallace's Selaginella (*Selaginella wallacei*), p. 36
Hansen's Selaginella (*Selaginella hanseni*), p. 36
Desert Selaginella (*Selaginella eremophila*), p. 36
Pygmy Selaginella (*Selaginella cinerascens*), p. 37
Oregon Selaginella (*Selaginella oregana*), p. 37
Alpine Selaginella (*Selaginella watsoni*), p. 37
Bluish Selaginella (*Selaginella asprella*), p. 37
Rocky Mt. Selaginella (*Selaginella densa* var. *scopulorum*),p. 37
Mojave Selaginella (*Selaginella leucobryoides*), p. 37

ADDER'S TONGUE FAMILY (Ophioglossaceae)

Western Adder's Tongue Fern (*Ophioglossum californicum*),p. 39

Moonwort (*Botrychium lunaria*), p. 40
Adder's Tongue Fern (*Ophioglossum vulgatum*), p. 40
Simple Grape Fern (*Botrychium simplex*), p. 42
California Grape Fern (*Botrychium silaifolium* var. *californicum;* in Munz, *B. multifidum* ssp. *silaifolium*), p. 43

FERN FAMILY (Polypodiaceae)

Golden Back Fern (*Pityrogramma triangularis*), p 47
Giant Chain Fern (*Woodwardia fimbriata*). p. 48
Western Spleenwort (*Asplenium vespertinum*), p. 48
Spleenwort (*Asplenium viride*), p. 48
Spleenwort (*Asplenium septentrionale*), p. 48
Lady Fern (*Athyrium filix-femina* var. *californicum*), p. 49
Leather-Leaf (*Polypodium scouleri*), p. 58
Licorice Fern (*Polypodium glycyrrhiza*), p. 58
California Polypody (*Polypodium californicum*), p. 58
Western Polypody (*Polypodium vulgare* var. *columbianum*), p. 60
Holly Fern (*Polystichum lonchitis*), p. 62
Sword Fern (*Polystichum munitum*), p. 62
Shasta Fern (*Polystichum lemmonii*), p. 62
Eaton's Shield Fern (*Polystichum scopulinum*), p. 63
California Shield Fern (*Polystichum californicum*), p. 63
Dudley's Shield Fern (*Polystichum dudleyi*), p. 65
Spreading Wood Fern (*Dryopteris dilatata*), p. 66
Coastal Wood Fern (*Dryopteris arguta*), p. 68
Male Fern (*Dryopteris filix-mas*), p. 68
Sierra Water Fern (*Dryopteris oregana;* in Munz, *Lastrea oregana*), p. 69
Downy Wood Fern (*Dryopteris feei;* in Munz, *Lastrea augescens*), p. 69
Oregon Woodsia (*Woodsia oregana*), p. 69
Rocky Mountain Woodsia (*Woodsia scopulina*), p. 70
Brittle Bladder Fern (*Cystopteris fragilis*), p. 71
Deer Fern (*Struthiopteris spicant;* in Munz, *Blechnum spicant*), p. 73
American Rock Brake or Parsley Fern (*Cryptogramma acrostichoides*), p. 74
Western Bracken (*Pteridium aquilinum* var. *lanuqinosum*), p. 74
Five Finger Fern (*Adiantum pedatum* var. *aleuticum*), p. 76
Common Maiden-Hair (*Adiantum capillus-veneris*), p. 76
California Maiden-Hair (*Adiantum jordani*), p. 78
Bridges' Cliff-Brake (*Pellaea bridgesii*), p. 79
Brewer's Cliff-Brake (*Pellaea breweri*), p. 79
Coffee Fern (*Pellaea andromedaefolia*), p. 79
Bird's-Foot Cliff-Brake (*Pellaea mucronata*), p. 79

Desert Cliff-Brake (*Pellaea compacta*), p. 81

Sierra Cliff-Brake (*Pellaea brachytera*), p. 85

California Lace Fern (*Cheilanthes californica;* in Munz, *Aspidotis californica*), p. 85

Indian's Dream or Oregon Cliff-Brake (*Cheilanthes siliquosa;* in Munz, *Onychium densum*), p. 85

Mrs. Cooper's Lip Fern (*Cheilanthes cooperae*), p. 85

Lace Fern (*Cheilanthes gracillima*), p. 85

Coville's Lip Fern (*Cheilanthes covillei*), p. 86

Viscid Lip Fern (*Cheilanthes viscida*), p. 87

Slender Lip Fern (*Cheilanthes feei*), p. 87

Fibrillace Lip Fern (*Cheilanthes fibrillosa*), p. 87

Parish Lip Fern (*Cheilanthes parishii*), p. 87

Cleveland's Lip Fern (*Cheilanthes clevelandii*), p. 87

Coastal Lip Fern (*Cheilanthes intertexta*), p. 87

Wooton's Lip Fern (*Cheilanthes wootonii*), p. 87

Jones' Cloak Fern (*Notholaena jonesii;* in Munz, *Cheilanthes jonesii*), p. 88

Parry's Cloak Fern (*Notholaena parryi;* in Munz, *Cheilanthes parryi*), p. 88

Cotton Fern (*Notholaena newberryi;* in Munz, *Cheilanthes newberryi*), p. 88

Cloak Fern (*Notholaena cochisensis;* in Munz, *Cheilanthes sinuata* var. *cochisensis*), p. 88

California Cloak Fern (*Notholaena californica;* in Munz, *Aleuritopteris cretacea*), p. 88